CUMBRIAN

IN THE

TWENTIETH CENTURY

Edited by

David J. Clarke & Stephen M. Hewitt

Transactions of the Carlisle Natural History Society
Vol XII (Centenary volume) 1996

Published with financial support from
Carlisle City Council
English Nature
British Ecological Society
Cumbria County Council
Botanical Society of the British Isles

Published and distributed by Carlisle Natural History Society
c/o Tullie House Museum & Art Gallery, Castle St, Carlisle CA3 8TP
tel: 01228-34781; fax: 01228-810249

Printed and bound in Great Britain by
Athenaeum Press Ltd, Gateshead, Tyne & Wear
Camera-ready text by Jeremy Roberts
Cover design by The Design House Group, Carlisle

© Carlisle Natural History Society and the authors, 1996

ISBN 0 9525252 0 8

Contents

Foreword by David Bellamy

Editors' introduction

Changes in Cumbria's flora in the 20th century 5
Geoffrey Halliday

Cumbrian butterflies since 1893 15
Geoff Naylor

Dragonflies in Cumbria – a centenary review 27
David Clarke

Beetles and beetle recording in Cumbria 39
Roger Key

Birds of prey in Cumbria 51
Geoffrey Horne

Mammals in Cumbria – a centenary review 77
John Webster

Wildlife and its conservation in Cumbria 89
Derek Ratcliffe

Carlisle Natural History Society – the first hundred years 107
Stephen Hewitt

Ernest Blezard (1902-1970) 127
Derek Ratcliffe

Contributors 133

FOREWORD

Over the past century, Natural History Societies abounded across Britain's biodiverse and pleasant land, each one allowing members, young and old, to rub shoulders with the geologists, botanists, entomologists, herpetologists, mammal, bird, snail, fungi, algae, moss, liverwort and lichen people of their day – each an expert (or rapidly developing expert) in their own right. Together they laid down a record of national biodiversity without equal, a record that now allows us to look back with envy, scientifically assess the ravages of our own times and lay real plans for the future.

This is a record that celebrates 100 years of painstaking work by one such Society, a Society which trod and recorded the holy ground of the English Lakes and its environs – in the words of Rawnsley (founder of the National Trust) "Nature's own English University". Back in those days, every University worth its salt had a thriving Natural History Society, but few with such a varied and important an estate as Cumbria, the landscapes of which saw the birth of the conservation movement, and gave strength to real campaigners who stopped what some, even in those days, would have called 'sustainable development' in its tracks.

Here is the record of our times. Though written by experts (and they *are* the real experts), it is redolent with everything Cumbria stands for – England's Green and Pleasant Land, brim full of flowers, butterflies, birds, mammals and people. The latter spilling out from the dark mill (and once boom) towns of Lancashire, and now flying in from around the nation and the world. Come read these pages, soar with Golden, or was it White-tailed, Eagle high over the crags, look down on Golden hosts of *Narcissus pseudonarcissus* and the tourists that still could love them to death. Celebrate the call of Skylarks, watch Trout hide amongst *Potamogeton* and *Najas* unsullied by the noise of power boats, and climb with the spirit of Christian Bonington, (though mind your boots), to the eyries where Alpine Mouse-ear and Spignel once flourished.

All thinking people know that the heart of this truly amazing county should be a World Heritage Site – this living record proves the case again and again. It also proves that with the knowledge it contains and with a little bit of care, vision and a lot of hard work Rawnsley's 'University' can, no *will*, be returned to its former glory – to enthral, educate, and yes entertain the millions of 'students' who come to drink from its extra-mural excellence.

There are immense problems. Amongst these are eutrophication – that is, the enrichment of soils and waters with phosphate and nitrate from agriculture, motor cars and human excrement; overgrazing by sheep and rabbits; visitor pressures of all sorts – boots, mountain bikes, 'off-roads', outboards and even light.

Yet there is one other factor that is both problem and solution: each year more than 15,000,000 tourists make use of the resource that is Cumbria. Sadly, I have to agree that few of them come to see the wild plants and animals, but they don't come to see erosion, litter, traffic cones, polluted waterways, square blocks of exotic conifers, barbed wire and monocultures. However, they do employ many locals and make profit for locals and absentees alike.

This book celebrates the fact that there is an answer. It shows that without the age-old land practices that nurtured the biodiversity of the place, the cost of management escalates and the cry for development to pay for its upkeep again raises its ugly head. At the moment, this management is carried out by the farmers, landowners and NGOs like the National Trust and the Cumbria Wildlife Trust. These diverse groups need all the help they can get, from Government, the public and from those who profit from the resource.

This is an amazingly important book, a Natural History of our times, buy it, cherish it, read all about it.

David Bellamy
Bedburn
June 1996

EDITORS' INTRODUCTION

This volume was planned to mark the centenary of the Carlisle Natural History Society, and is based on the Centenary Conference which was held at Tullie House Museum on 25 September 1993.

The post-Conference interlude provided an opportunity to expand and update the papers, and to include additional information on sources and other background. Derek Ratcliffe's appreciation of the noted Carlisle naturalist Ernest Blezard was not available for the Conference and has proved a fitting and opportune addition to the record of the Society's history, with which this book concludes.

Partly for financial reasons, the finished publication has taken longer in the making than we originally envisaged. In any event, it would have been quite impossible without the substantial support from the various funding bodies listed on the title page. We thank them all profusely.

Many individuals have made special contributions, and in particular we wish to extend thanks to the authors of the various papers. Our gratitude goes also to Jeremy Roberts for generously donating his desk-top publishing skills; to Jeremy and Margaret Roberts for much other assistance at proof stage; to Margaret Robinson of the Tullie House staff for help with various word-processing tasks.

We are especially pleased that two leading figures in the world of natural history and wildlife conservation are associated with this book. Derek Ratcliffe, eminent field naturalist, research ecologist, author and former Nature Conservancy Council Chief Scientist, has played a special part in British natural history. As well as contributing directly to this book, his particular associations with Carlisle and Cumbria have enabled him to offer much other valuable comment. He joined Carlisle Natural History Society in 1944. David Bellamy, botanist and University Professor turned activist and popular interpreter, has become one of the best-known public figures in the world of wildlife conservation. Currently national President of the Wildlife Trusts, he too has a special affinity with, and affection for, northern England; we value his endorsement of our work.

We hope that in a modest way, the contents of this book will not only provide a unique record of a very special county, but also serve as a tribute to all the naturalists, professional and amateur alike, who have laboured hard, and too often unsung, to reveal the true picture of events in the natural world over a century of unprecedented change.

David Clarke and **Stephen Hewitt (Editors)**
Tullie House Museum & Art Gallery, Carlisle
September 1996

CHANGES IN CUMBRIA'S FLORA IN THE 20th CENTURY

Geoffrey Halliday

The centenary of this Society is an appropriate time to review changes in the flora of Cumbria over the last century, since it is almost a century since the publication of Hodgson's *Flora of Cumberland* in 1898. In the absence of any acceptable list for that period for the modern county of Cumbria, his *Flora* provides a useful base-line to compare with Cumberland records accumulated over the last twenty years in connection with the Flora of Cumbria Survey. If dubious records and microspecies of hawkweeds *(Hieracium)*, dandelions *(Taraxacum)*, brambles *(Rubus)* and eyebrights *(Euphrasia)* are excluded, Hodgson lists 1145 plants which are currently given specific rank. In addition to native species, this figure includes ones which are doubtfully native in Cumberland, as well as garden-escapes and other aliens. He lists 207 species in these last categories. 86 are presumed garden-escapes or planted and 121 are best regarded as casuals, of which only 39 have been recorded during the recent Survey. Most of the species which have disappeared are Hodgson's 'waifs' of the ballast tips and dock areas of the then thriving west Cumberland ports, chiefly Workington, Silloth and Maryport, which were such a happy hunting ground for botanists from the 1870s until well into the early decades of this century. Hodgson's figure of 207 species can be substantially increased by those noted by J. Leitch and E.J. Glaister (mainly in the Silloth area) in the latter part of last century, and by T.S. Johnstone in the early years of the present century around Carlisle – particularly on the gravel beds of the River Eden. The picture is similar in south Cumbria, where W.H. Pearsall and D. Lumb recorded numerous aliens from the steelworks and ports of Barrow and Askam-in-Furness.

These figures can be compared with the recent Survey records for Cumberland. We have recorded 1191 species of which 290 are aliens, including established garden-escapes and throw-outs. This massive increase (140%) in aliens is hardly surprising given the ever-increasing popularity of gardening and the wide variety of species now grown, as well as the deplorable habit of dumping surplus material, especially of excessively aggressive species, on roadside verges and in ditches where, not surprisingly, many persist. The similarity of the totals for the beginning and end of the century is surprising and indicates that the increase in the number of aliens of garden origin and of 15 native species more than offsets the loss of the many 19th century casuals and 56 native species.

These figures of course relate to Cumberland and would be substantially boosted, particularly in respect of garden-escapes, by the inclusion of Furness and Westmorland.

The above is a crude comparison of the number of species. I shall now consider examples of individual losses and gains in the county as a whole.

Losses

No-one can browse through Hodgson's Flora without being struck by the vast number of weeds from the ballast and industrial sites of the west coast ports. Many of these are natives of southern Europe but some are from even further afield, such as species of *Verbesina, Ximenesia* and *Bowlesia* from America. Most failed to survive the demise of these ports and have not been seen since. This is true, for example, of three species of clover *(Trifolium* spp.*)* four knapweeds *(Centaurea* spp.*)* and four brome grasses *(Bromus* spp.*)*

Major changes in agriculture, particularly since the last war, have had a major impact on the county's flora. Of the 56 native species which have been recorded in the past from Cumberland but not seen during the Survey, 31 (55%) disappeared between 1930 and 1970. Similarly, in the whole of Cumbria, 16 (50%) of the 32 native species previously recorded, but not seen during the Survey years, disappeared within that period. There has been a dramatic decline in traditional hay meadows, which have been replaced by 'improved' grassland and leys either for hay or, especially in recent years, for silage. Such changes have resulted in the loss of the Burnt Orchid *(Orchis ustulata)* and the increasing rarity in the dales of the Globe Flower *(Trollius europaeus)* and the Melancholy Thistle *(Cirsium heterophyllum).* The Bird's-eye Primrose *(Primula farinosa)* is now very rare in the limestone district between the Lake District and Wigton where Hodgson said it was abundant over several miles. There are 23 localities given in the old Floras for the Small-white Orchid *(Pseudorchis albida)* but there are now only three; fortunately it still persists in a number of sites in the former Yorkshire sector of Cumbria. During the same period, with the increasing use of herbicides there was a corresponding decline in arable weeds resulting in the disappearance from Cumbria of Corncockle *(Agrostemma githago)*, Shepherd's Needle *(Scandix pecten-veneris)* and the cornsalads *Valerianella dentata* and *V. rimosa,* and of Knotted Hedge-parsley *(Torilis nodosa)* from Cumberland.

Moorland drainage and afforestation have also taken their toll. Chickweed Wintergreen *(Trientalis europaea)*, once known from several sites in its stronghold of the Bewcastle Fells, is now very rare and the club-moss *Lycopodiella inundata,* recorded by Hodgson from Wastwater, Borrowdale and Thirlmere, is now reduced to just three sites, one of which is outside the Lake District. Drainage is no doubt also responsible for the disappearance of Rigid Hornwort *(Ceratophyllum submersum)* at Monkhill Lough, Carlisle, one of its only two sites in Cumbria, and of Marsh Gentian *(Gentiana pneumonanthe)* from its two nineteenth century Cumberland sites.

The eutrophication of lakes and rivers, by fertiliser run-off and sewage effluent, has had a dramatic effect. It may well have been the cause of the extinction of *Hydrilla verticillata* (first seen in 1914) from Esthwaite Water, its only British locality, and the apparent decrease there of *Najas flexilis* in its only English site. Most striking has been the widespread loss of pondweeds *(Potamogeton* spp.*)* from the tarns and particularly the lakes.

There have probably been very few changes in status of the rarer montane species. Provided their cliff habitats are out of reach of the sheep, the major threats come from rock-falls and the inevitable fluctuations in numbers of small, isolated populations. The wintergreen *Pyrola media* has become extinct and rock falls appear to have been responsible for the loss in the Lake District of a site for the Holly-fern *(Polystichum lonchitis)*, both sites of the Limestone Fern *(Gymnocarpium robertianum)* and possibly of the only sites for the Alpine Penny-cress *(Thlaspi cærulescens)* and the grass *Phleum alpinum*. The fern *Woodsia ilvensis* is now reduced to a single site. Neither Alpine Mouse-ear *(Cerastium alpinum)* nor Interrupted Clubmoss *(Lycopodium annotinum)* have been seen in Cumberland this century, nor Spignel *(Meum athamanticum)* which used to occur in the Keswick area. During the two decades preceding the Survey, Rigid Buckler-fern *(Dryopteris submontana)* vanished from its only Lake District site at Honister, as a result of a rock fall, and the lady's-mantle *Alchemilla monticola* from its only Cumbrian site on roadside verges on the Durham border. Shrubby Cinquefoil *(Potentilla fruticosa)*, lost from Helvellyn, and Mountain Avens *(Dryas octopetala)*, lost from the Cross Fell area, have, however, each been discovered at a new site in the Lake District. Grazing probably eliminated the ragwort *Tephroseris integrifolia* from the fells above Brough, where it has not been seen since the 1920s. Fortunately, collecting seems no longer to be a serious problem. It probably resulted in the disappearance last century of the fern *Cystopteris montana* from Helvellyn, its only English locality, the Lady's-slipper Orchid *(Cypripedium calceolus)*, and the apparent restriction of sporophytes of the Killarney Fern *(Trichomanes speciosum)* to a single site. A remarkable recent discovery was of gametophytes of this fern at a few places in the central Lake District.

Gains

These fall into four general categories:

1. **Weeds**

    These are the casuals of roadsides, landscaped and re-seeded ground, urban areas and waste sites. They include Oxford Ragwort *(Senecio squalidus)*, first recorded in 1935 and now well established around Carlisle, the west coast towns, and Barrow-in-Furness, and the American Willowherb *(Epilobium ciliatum)*, very widely distributed although unknown before *ca* 1969. The extremely rare southern broomrape *Orobanche purpurea* was discovered at Maryport in *ca* 1983 and there are Survey records of the hawk's-beards *Crepis tectorum* and *C. setosa*, the grass *Bromopsis inermis*, as well as *Amaranthus retroflexus, Bassia scoparia, Portulaca oleracea* and the decreasing southern weed Venus's-looking-glass *(Legousia hybrida)*, which was found in 1990 on a motorway bridge. Tree Mallow *(Lavatera arborea)*, Lovage *(Levisticum officinale)*, the mugwort *Artemisia stellerana* and *Lupinus arboreus* have all appeared on the west coast, but the first two failed to persist. Although known from only three pre-Survey sites, the North American grass *Hordeum jubatum* became very noticeable in the early years of the Survey along the saline verges of motorways and trunk roads although it appears to have declined somewhat in recent years.

2. **Aquatics**

   Cumbria has shared in the spectacular national spread of the North American waterweeds *Elodea canadensis* (first seen in the county in 1874), and recently of *E. nuttallii* (1976) which seems to be replacing it, at least in several of the lakes. The figwort *Scrophularia umbrosa* (1924), **fig. 1**, is now well distributed along the whole length of the River Eden and even more at home throughout the county by rivers, lakes and canals is the Policeman's Helmet *(Impatiens glandulifera)* (1921), **fig. 2**. Gardeners and garden centres are no doubt responsible for the vegetative spread not only of the *Elodea* spp. but also the recent appearance of *Lagarosiphon major*, the Water Fern *(Azolla filiculoides)*, the water-lily *Nymphoides peltata* and *Crassula helmsii*, perhaps also the appearance in inland fish ponds in the east of the water-milfoil *Myriophyllum spicatum*. The *Crassula*, a native of New Zealand and Australia, is now an aggressive weed in southern England and its status in Cumbria will need to be watched carefully, particularly following its discovery in Derwent Water. The North American Pitcherplant *(Sarracenia purpurea)* is well established on mossland by the Solway Firth and at a number of wetland sites between Windermere and Coniston where it is known to have been planted. Another North American species is the elegant loosestrife *Lysimachia terrestris* (1884), now a characteristic feature of tarn and lakeside marsh communities around the southern half of Windermere.

3. **Garden escapes**

   The dumping of excess garden material and the movement of soil containing rootstocks and rhizomes is increasingly noticeable throughout the lowlands. Especially conspicuous are the resulting dense thickets of the Japanese Knotweed *(Fallopia japonica)* (1930s), **fig. 3**, and to a lesser extent, the related *F. sachalinensis* and *Polygonum polystachyum*. The eye-catching butterburs *Petasites japonicus, P. albus* and *P. fragrans* have probably originated in this way although some roadside colonies are known to have been planted. Also conspicuous are the deep yellow loosestrife *Lysimachia* *punctata*, the deep purple-flowered *Geranium* × *magnificum*, the pink-flowered *G. endressii* and the pale blue sow-thistle *Cicerbita macrophylla* (1915). Other newcomers are *Tolmiea menziesii* and *Tellima grandiflora*. The showy, purple Pyrenean cress *Cardamine raphanifolia* is frequent in wet places in the Grasmere/Ambleside area and Monk's-hood *(Aconitum napellus)* and Jacob's-ladder *(Polemonium cœruleum)* are well distributed although not common. Bird-sown species of *Cotoneaster* are particularly frequent in the Arnside area and another species which has needed little in the way of help from man is the now almost ubiquitous New Zealand Willowherb *(Epilobium brunnescens)* (ca 1923), **fig. 4**. The attractive, early-flowering speedwell *Veronica filiformis* (1914) is now common in the lowlands in lawns and on roadside verges and this despite its inability to set seed. There have been a number of recent records of the variable mallow-like *Sidalcea*, several of the popular garden poppy *Papaver orientale* and a single remarkable occurrence on a limestone beckside east of Sedbergh of the Pyrenean rock garden plant *Ramonda myconi*. There is no mention in Hodgson's Flora, nor in Wilson's (1938) Westmorland Flora, of the North American goldenrods *Solidago gigantea* and *S. canadensis* which are spreading fast, nor, remarkably, of *Rhododendron ponticum*, now a pernicious weed in many of our woodlands. Other shrubs which appear to be spreading, although less spectacularly, are *Lonicera nitida*, Lilac *(Syringa vulgaris)* and *Buddleja davidii*. A surprising find in the east of the county was the willow *Salix udensis*, the only English record.

As a result of 20th century commercial forestry, Sitka Spruce *(Picea sitchensis)* and Lodgepole Pine *(Pinus contorta)* are now familiar species in the county and Larch *(Larix decidua)* is so common that it comes as a surprise to find that Hodgson fails to mention it, although it has been present since the late 18th century. Also increasingly conspicuous are the roadside plantings of poplars, especially the Balsam Poplar *(Populus trichocarpa)*, the dogwood *Cornus sericea*, the Wayfaring-tree *(Viburnum lantana)* and, occasionally, Sweet-briar *(Rosa rubiginosa)*, as well as daffodil hybrids on roundabouts.

4. **Native species**

Excluding microspecies and casuals, 47 species regarded as native in the county were not discovered until this century; the corresponding figure for Cumberland is 42. Pre-Survey records include the following: the helleborines *Epipactis leptochila* var. *dunensis* and *E. phyllanthes*, the wintergreen *Pyrola rotundifolia* and Coralroot Orchid *(Corallorhiza trifida)* were all discovered on the dune-slacks of the south-west coast during the inter-war years. The south-west coast also produced the sedge-like *Isolepis cernua* and Smooth Cat's-ear *(Hypochœris glabra)*. Particularly interesting were the finds of the sedge *Carex elongata* in the Lake District and south

Furness, and of *C. ericetorum* and Teesdale Violet *(Viola rupestris)* in the Arnside area and the limestones of north Westmorland. The Asby/Orton limestone also yielded two sites for the cress *Hornungia petraea* and, nearby, sites for the 'Teesdale' milkwort *Polygala amarella*. The outlying Cumbrian sites of the southern gorse *Ulex minor* near Carlisle were not discovered until the 1940s and more were added during the Survey. Only 13 native species have been added during the Survey and not surprisingly half are members of critical groups, for example *Alchemilla glaucescens*, a lady's-mantle of the Craven area of Yorkshire, the grass *Poa angustifolia* (but possibly introduced), the glasswort *Salicornia fragilis*, and the chickweed *Stellaria pallida*. Other additions include the quillwort *Isoetes echinospora* at three sites, the very rare spike-rush *Eleocharis austriaca* in the upper River Irthing, the tasselweed *Ruppia cirrhosa* near Askam-in-Furness, one of its only two extant sites on the west coast of Britain, and the sedges *Carex divulsa* near Barrow-in-Furness and *C. maritima* at Humphrey Head. The last two records are respectively the most northerly and southerly ones on the west coast of Britain; unfortunately *Carex maritima* failed to survive coastal defence operations. Survey additions for Cumberland include the cord-grass *Spartina anglica* at Ravenglass (but not yet on the south side of the Solway) and, most surprisingly, the Bird's-foot Sedge *(Carex ornithopoda)* by the River Eden near Armathwaite. Arguably the most important find of the Survey was of the hawk's-beard *Crepis præmorsa* in a hay-field in north Westmorland and new to western Europe.

Particularly as a result of the current detailed Survey, there has been an appreciable increase in the number of known sites of species previously regarded as rare or very rare. These include *Saxifraga hirculus* in the Pennines, Alpine Cinquefoil *(Potentilla crantzii)* both in the Lake District and the Pennines, and the predominantly eastern cottongrass *Eriophorum latifolium*. Before R. Stokoe's systematic investigation of Cumbrian lakes and tarns during the early years of the Survey, the diminutive waterweed *Elatine hexandra* was known with certainty only from Thurstonfield Lough yet he succeeded in finding it in eight of the lakes and twelve tarns. There has been a puzzling increase in the records of another waterweed *Zannichellia palustris*. This is now widespread although Hodgson cites only five localities and Wilson two; none are from the Eden valley where it is now quite frequent. In some cases, as for example with the Bog Orchid *(Hammarbya paludosa)*, the number of new sites seems to have kept pace with the loss of old ones.

Not surprisingly, as a result of increasing taxonomic knowledge the number of recorded microspecies and segregates is continually increasing. The bulk of the 40 Cumbrian species of bramble, the 130 dandelions and the 60 hawk-weeds were first described during this century. The same is true of our three

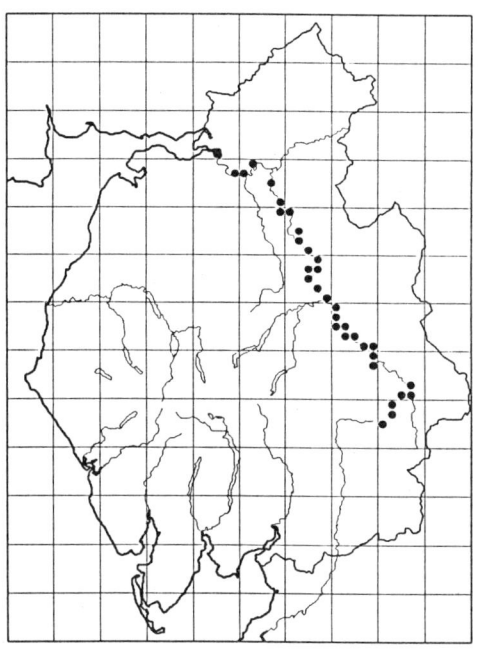

fig. 1:
Green Figwort
Scrophularia umbrosa
First recorded in Cumbria in 1924

fig. 2:
Policeman's Helmet
Impatiens glandulifera
First recorded in Cumbria in 1921

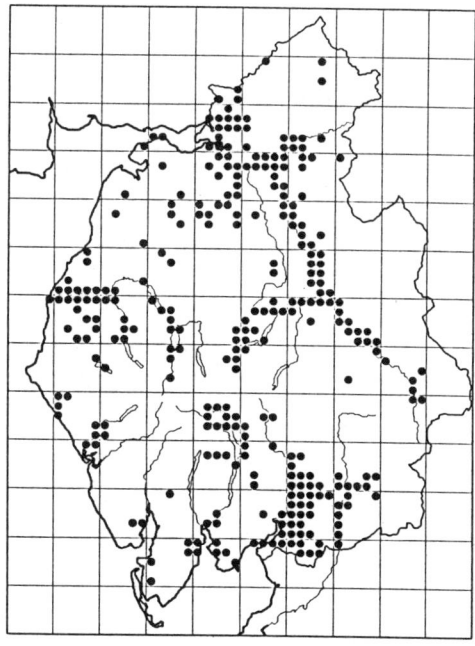

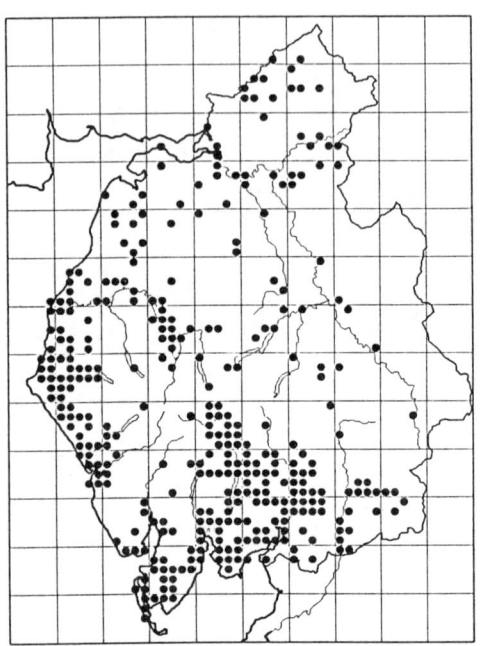

fig. 3:
Japanese Knotweed
Fallopia japonica
First recorded in Cumbria in the 1930s

fig. 4:
New Zealand Willowherb
Epilobium brunnescens
First recorded in Cumbria *ca* 1923

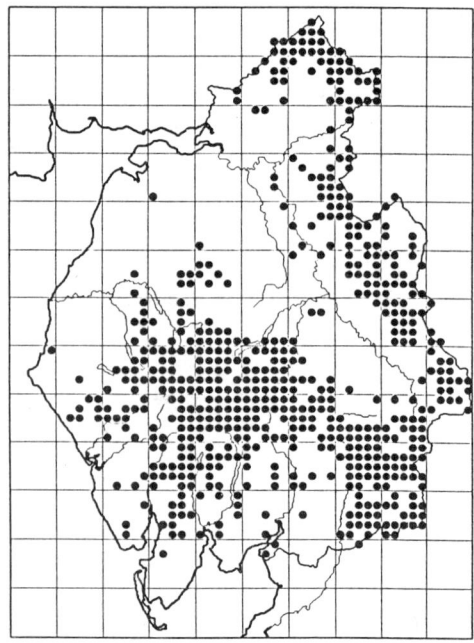

scarce Lake District eyebrights: *Euphrasia frigida*, *E. rivularis* and *E. ostenfeldii*.

Conclusion

The increase in the number of new native species recorded over the last century is the result of increased taxonomic knowledge combined with more intensive searching. Relatively few species have been lost but the rare species and their habitats are getting rarer. Cumbria is still an extraordinarily rich county but this is no reason for complacency. We have a national responsibility to protect what still remains. Although it makes heavy demands on our resources, we must intensify our efforts to identify those species and habitats which are at risk and act before it is too late.

References

Halliday, G., (in press), *A Flora of Cumbria*.
Hodgson, W., (1898), *Flora of Cumberland*. Carlisle: Meals.
Wilson, A., (1938), *The Flora of Westmorland*. Arbroath: Buncle.

Editors' note: English and scientific names of plants follow *New Flora of the British Isles,* 1991, C.A. Stace (C.U.P.)

CUMBRIAN BUTTERFLIES SINCE 1893

Geoff Naylor

The popularity of butterflies with Victorian and Edwardian collector-naturalists has ensured that the early years of the review period were particularly well documented. Indeed, the years 1893 to 1933 might be described as the 'Golden Age' of Cumbrian lepidopterists. This period saw the peak of the activities of G.B. Routledge, F.H. Day, H.A. Beadle and Harry Britten in the north, and A.E. Wright and Dr R.C. Lowther in the south. Their predecessors T.C. Heysham, J.B. Hodgkinson and George Dawson had already laid important foundations of knowledge of the local fauna. (F.H. Day in fact continued his entomological interests until about 1960, but concentrated on other insect groups in later years.)

In the middle years of the century, G.A.K. Hervey and J.H. Vine Hall also made significant contributions to work which has been carried forward by N.L. Birkett, John Heath, E.F. Hancock, D.W. Kydd and W.R. Laidler in particular. John Heath did much recording in the Furness area before moving to Monks Wood to start the national insect mapping scheme. His initiatives were the catalyst for the post-war resurgence of recording, which has continued to gather momentum ever since. Neville Birkett has been active for over sixty years and contributed many papers on Lepidoptera.

Around 100 years ago some 42 species of butterfly were regularly recorded in the county. To these may be added some more doubtful old records, which will be referred to later. In the late nineteenth and early twentieth centuries most records are referable to the then county of Cumberland (vice-county 70). Information about neighbouring Westmorland and Lancashire North of the Sands (or Furness as it is alternatively known) (vc. 69) is more sporadic. The latter area has, for example, been covered by the Lancashire and Cheshire Entomological Society. There is virtually no historic information regarding the Dent and Sedbergh areas which were then part of Yorkshire (vc. 65).

Today, the number of species which regularly occur in the county is around 37 – an 11% loss. This simple fact conceals significant changes in the distribution and abundance of many species over the century.

Pre-1893 and doubtful records

The Small Skipper has been reported three times from the Carlisle area (the last in 1950), but none of these can be confirmed. There is one old reference to the Silver-spotted Skipper, from 3 sites west of Carlisle (Dawson 1879), but in view of the present distribution and habitat of this species, it is unlikely that this was correct. A Swallowtail was recorded at Gilsland (NY66) by Hodgkinson in the last century, and is listed as 'doubtful' by Routledge (1909). Continental Swallowtails do occur infrequently as migrants to Britain, so this record should perhaps not be completely discounted. Routledge also refers to the Chalkhill Blue, as being *'taken not uncommonly at Grange . . .'* and even to a former colony reportedly from [Mun]Grisedale, under Saddleback. As the nearest sites (including extinct colonies) are in Lincolnshire, such records now seem hard to credit and the writer is not aware of any voucher specimens. Also among Routledge's 'doubtfuls', Queen-of-Spain Fritillaries had been claimed in September 1833, and a White Admiral in 1859 – both at Carlisle. As an irregular migrant, the former species cannot be completely ruled out. Finally, and much more recently, there is a possible sighting of a White-letter Hairstreak at Kendal in August 1984.

Extinct species

Wood Whites were once known from the south of the county, the Lake District, and several sites near Carlisle, including Baron Wood and Newbiggin Woods. There have been no records since 1920 (but see page 21). This was also the last recorded year for the Brown Hairstreak. The latter was mainly known from locations in the Grange/Witherslack area. (There was also a possible record from Baron Wood.) A speciality of some south Cumbria Mosses, the Silver-studded Blue was finally brought to extinction by the disastrous fire at Meathop Moss (SD48) in 1940, after previous losses at other sites. The disappearance of this isolated northern population was particularly unfortunate because the form was a unique one, regarded by some as a subspecies *Plebejus argus masseyi*. The Large Tortoiseshell has not been recorded in Cumbria since 1945, and is now believed extinct in Britain as a breeding species. There have only been a handful of records of the Grizzled Skipper (from the Arnside area), and none since 1929.

Species showing little change in status

The Large, Small and Green-veined Whites, Orange-tip, Small Pearl-bordered Fritillary, Small Tortoiseshell, Meadow Brown, Large Heath, Small Heath and Common Blue have shown little change in status over the century. The Ringlet population, although steady, shows some unusual features, being widespread in the former county of Cumberland, but almost unknown in Westmorland – apart from a curious isolated cluster in the Smardale area (**fig. 1**).

Also in this category are those species which have always been less widespread, but whose populations have remained fairly stable: High Brown Fritillary, Green Hairstreak, Northern Brown Argus, Mountain Ringlet and Scotch Argus. The High Brown occurs only in south Cumbria but probably has its most stable British population there; Green Hairstreak shows a few 'gaps' in its county ditsribution and may be an under-recorded species; Northern Brown Argus is restricted to Smardale and sites south of Kendal, but does not appear endangered; Mountain Ringlet and Scotch Argus have their only English sites in Cumbria, both with healthy populations.

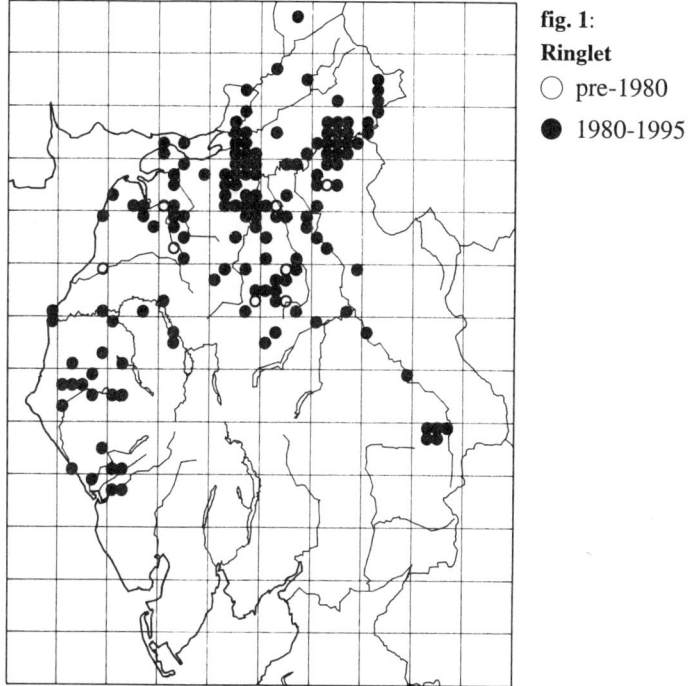

fig. 1:
Ringlet
○ pre-1980
● 1980-1995

Species which have significantly declined

Seven species seem to have contracted their ranges since 1893, but are still locally common.

The Dingy Skipper may have suffered losses, but has recently been found at some new sites. The Brimstone seems a 'sensitive' species. Its small population is near to its northern limit in Britain and tends to 'crash' in some years. The Purple Hairstreak is not uncommon in southern Lakeland, but records from the north of the county have been few in recent years – 1995 being an exception. It is, however, a difficult species to locate. The Holly Blue is now only regular south of Windermere (where it has actually recovered since being scarce in the 1960s), with only scattered reports from more northerly areas. Peacocks almost disappeared in 1986-7, but despite an improvement in the south of the county have until very recently shown only a fitful recovery in the north. The Wall Brown is now found mainly in coastal areas, apart from Carlisle and the Solway plain. The Grayling has become even more strictly coastal, with a mere 4 or 5 inland sites at present. The latter include limestone scars in the south of the county and disused railway sidings at Carlisle.

Scarce and restricted species

Four species are now severely restricted, have greatly declined and are thus potentially vulnerable. The Small Blue is limited to only 2 or 3 coastal areas. Formerly quite common in parts of NE Cumbria, it has not been recorded at any of these sites since the 1940s, when the management of railway cuttings in the days of steam was especially suited to its foodplant, Kidney Vetch. The Duke-of-Burgundy is found at only a dozen or so sites, all south of Kendal. Unlike most of the rarities, the Marsh Fritillary is restricted to north Cumbria. It is now found in only 2 areas, one of which has two or three sites. Formerly it was rather more widespread. In 1930, six colonies were known within 7 miles of Carlisle (Ford & Ford, 1930). All of these are now extinct, including the famous Orton Moss site, where the Fords studied genetic variation in this species. The Pearl-bordered Fritillary seems to have retreated to the extreme south of the county (**fig. 2**). It is sometimes reported through misidentifications of the very similar Small Pearl-bordered Fritillary, and may also be overlooked for the same reasons. Some of the former localities for this species may therefore repay further checking.

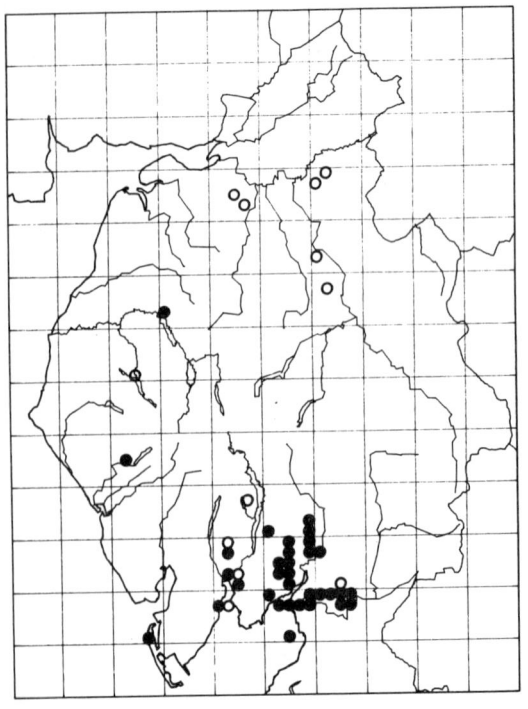

fig. 2:
Pearl-bordered Fritillary
○ pre-1980
● 1980-1995

Migrant species

There have been unconfirmed reports of the Pale Clouded Yellow in some of the 'Clouded Yellow years' and 14 were reported in Westmorland and Furness in 1947 (Heath & Emmet, 1989). Invading Clouded Yellows had a spectacular year nationally in 1992: 172 records were received for Cumbria (Naylor, 1993), there having been 102 over the previous one hundred years! A few were seen in 1994. Previous 'good years' for this species were 1983, 1980, 1947 and 1877. The Camberwell Beauty, a much scarcer migrant, has been noted in eleven seasons over the review period. Four individuals were seen in 1976-7, 13 in 1995 and a presumed overwintering individual in April 1996. An American species, the Monarch was added to the county list quite recently, with single occurrences in 1980 and 1981, and four in 1995. The Painted Lady and Red Admiral are common in some years, and usually widespread. Increasing observer numbers may have improved recent detection levels for all migrant species.

Species which are currently increasing

Although mostly present in small numbers, the following species may be increasing:

The Speckled Wood has a scattering of old records from Keswick, Witherslack and, oddly, Shap (specimen at Carlisle Museum). There is also a single recent report from Loweswater in 1982. Small numbers are now established in the Arnside and Witherslack areas, but undoubtedly owe their existence to re-introductions. Gatekeepers are common on and near the coast between Whitehaven and the Duddon estuary, at the north limit of their British range. There has been recent expansion eastwards in Furness, and northwards along the coast to beyond Whitehaven. After a long interval since previous records, Commas have spread into south Cumbria in small numbers from 1990 on. They were recorded near Keswick and Carlisle in the exceptional summer of 1995. The Silver-washed Fritillary now appears to be flourishing very locally in south Cumbria after a handful of previous records this century. Whether it has recolonised, been re-introduced or persisted naturally remains a matter for debate. The first seems unlikely given the isolated nature of the Cumbrian population.

Future prospects

Hopefully the Comma will now continue to increase its base in the county. The Small Skipper is now expanding northwards up the east coast of Britain. Conceivably it could reach Cumbria. The White-letter Hairstreak now has a colony very close to the south Cumbria border and must be an even stronger contender.

For differing reasons, five species in particular may prove to have been under-recorded: the Dingy Skipper, Holly Blue, Green Hairstreak, Purple Hairstreak and Pearl-bordered Fritillary. It is to be hoped that systematic searching will produce more records of all of these.

On a cautionary note, it is always worth remembering that, more than any other insect group, butterflies are and always have been the subject of introductions. Collectors, breeders and other enthusiasts, both deliberately and accidentally, are responsible. While the effects of such activities are usually marginal, they can never be disregarded, especially where rare or unusual species or records are concerned.

Concluding remarks

Increased interest and more intensive monitoring of butterflies in recent years make direct comparison with historic records difficult. However, there have clearly been substantial changes over the past century. Of the four species which have disappeared from the county, perhaps only the Silver-studded Blue's demise can be clearly related to human activity. Nonetheless, many of the remaining species have declined due to changes in land management practice. Many of the Small Blue and Marsh Fritillary losses can be directly linked to habitat loss and circumstantial evidence exists for a similar cause of decline in several other species. Human activities may also be beneficial – the resurgence of the Silver-washed Fritillary mirrors the reinstatement of rotation coppicing of the woodlands concerned.

Thus it is clear that we must accept responsibility for the fate of our local butterflies, be it extinction or recovery. Increased concern and knowledge of the factors affecting butterfly populations has resulted in recent initiatives to conserve threatened butterflies in the county. The parlous state of the Marsh Fritillary precipitated the formation of the Cumbria Marsh Fritillary Action Group which aims to reverse the decline through habitat enhancement and careful re-introduction. Similar projects operate for the High Brown Fritillary and Duke of Burgundy.

To date, improved data collection through the recording scheme operated by Bill Kydd (for Cumbria Naturalists' Union) and Tullie House Museum has enabled targetted surveys of vulnerable species such as the Small Blue, where Robert Park has recently provided precise information on the distribution of the species in north west Cumbria. Maureen Richards has coordinated a survey of the Mountain Ringlet in recent years. Gaps in current knowledge have also been revealed with several 'under-recorded' species warranting further attention. The Large Heath is one such species, with its status on many Cumbrian sites far from clear at a time when recent work in Northumberland suggests a decline in that county. Finally, as the distribution of species in the county is becoming resolved, monitoring of key colonies and sites is becoming increasingly valid and important.

The rising popularity of butterflies has led to the formation of a new national organisation devoted to protecting these insects – Butterfly Conservation. With increasing membership and resources, the contribution from this organisation to the conservation of our local insects is to be welcomed. They have been involved in many of the initiatives mentioned above.

Acknowledgements

Bill Kydd and Ted Hancock have kindly provided much information from south Cumbria – the former has collated county records on behalf of the Cumbria Naturalists' Union for many years. Butterfly Conservation and the Cumbria Wildlife Trust have provided useful assistance with the recording scheme. Brian Eversham of the Biological Records Centre at Monks Wood has enabled access to Cumbria records held on the national database. Thanks are also due to all who have sent field records to Carlisle Natural History Society, Cumbria Naturalists' Union and the Tullie House Museum Biological Records Centre.

Cumbrian butterflies: summary of current status (1995)

D – doubtfully recorded; E – extinct; R – resident; RM – regular immigrant; IM – irregular immigrant; RC – recent colonist.

Small Skipper
(Thymelicus sylvestris)
D. Not certainly recorded. Might eventually colonise.

Silver-spotted Skipper
(Hesperia comma)
D. One very doubtful old record.

Large Skipper
(Ochlodes venata)
R. Moderately common in some areas, mainly in coastal lowlands.

Dingy Skipper
(Erynnis tages)
R. Not common. A few strong colonies, and some recent discoveries, though few recent records in the north. Habitats include waste ground.

Grizzled Skipper
(Pyrgus malvae)
E. Last recorded 1929, return unlikely given present range.

Swallowtail
(Papilio machaon)
D. One old and uncertain record.

Wood White
(Leptidea sinapis)
?E. No records after 1920 until a few individuals were seen at a site in the south of the county in 1996. This is assumed to be a recent reintroduction.

Pale Clouded Yellow *(Colias hyale)*	IM. Several unconfirmed reports from 'Clouded Yellow years'.
Clouded Yellow *(Colias croceus)*	IM. Abundant in invasion years; a few records almost annually.
Brimstone *(Gonepteryx rhamni)*	R. Moderate numbers occur in south Cumbria; few records from further north, but recently recorded near Penrith.
Large White *(Pieris brassicae)*	R. & RM. Widespread and normally common.
Small White *(Pieris rapae)*	R. Widespread and common.
Green-veined White *(Pieris napi)*	R. Widespread and common.
Orange-tip *(Anthocaris cardamines)*	R. Frequent throughout – often on roadside verges.
Green Hairstreak *(Callophrys rubi)*	R. Widespread in suitable habitats, especially heaths with bilberry, sometimes in moderately upland sites.
Brown Hairstreak *(Thecla betulae)*	E. Extinct since 1920 and unlikely to re-establish.
Purple Hairstreak *(Quercusia quercus)*	R. Elusive. Not uncommon in south Lakeland, but scarcer in the north.
White-letter Hairstreak *(Satyrium w-album)*	D. A possible record in 1984, but this species could colonise from north Lancashire.
Small Copper *(Lycaena phlaeas)*	R. Widespread but erratic. Third broods sometimes yield good numbers.
Small Blue *(Cupido minimus)*	R. Formerly more widespread, now reduced to two or three colonies.
Silver-studded Blue *(Plebejus argus)*	E. Extinct by 1940; several former sites in the south and Lake District.

Northern Brown Argus *(Aricia artaxerxes)*	R. Several sites in south and occasionally seen further north.
Common Blue *(Polyommatus icarus)*	R. Widespread and common.
Chalk-hill Blue *(Lysandra coridon)*	D. A few old and uncertain 19th century records.
Holly Blue *(Celastrina argiolus)*	R. Generally scarce, but more frequent in the south. A few second broods are seen in most years in the south of the county.
Duke of Burgundy *(Hamearis lucina)*	R. Limestone districts from Kendal to Morecambe Bay support several strong colonies.
White Admiral *(Ladoga camilla)*	D. One uncertain 19th century record.
Red Admiral *(Vanessa atalanta)*	RM. Common in many years; occasional overwintering is suspected.
Painted Lady *(Cynthia cardui)*	RM. Scarce or even absent in some years.
Small Tortoiseshell *(Aglais urticae)*	R. Common and widespread.
Large Tortoiseshell *(Nymphalis polychloros)*	E. Very few records – none since 1945.
Camberwell Beauty *(Nymphalis antiopa)*	IM. Some 28 reports over 11 seasons. 1995 produced 13 of these.
Peacock *(Inachis io)*	R. Formerly common, but 'crashed' in 1986 and is now slowly returning.
Comma *(Polygonia c-album)*	RC. Very rare in past, recently noted in small numbers in south from 1990. Seen further north in 1995 and 1996.
Small Pearl-bordered Fritillary *(Boloria selene)*	R. Common in suitable habitats, though probably declining.

Pearl-bordered Fritillary *(Boloria euphrosyne)*	R. Now restricted to area between Kendal and Morecambe Bay. Formerly more widespread (including Carlisle area). Fig. 2, p. 18.
Queen of Spain Fritillary *(Argynnis lathonia)*	D. A 19th century record - possibly genuine.
High Brown Fritillary *(Argynnis adippe)*	R. Strong colonies in south Cumbria are nationally important. Occasional recent records further north require confirmation.
Dark Green Fritillary *(Argynnis aglaja)*	R. Probably declining. Scattered throughout county.
Silver-washed Fritillary *(Argynnis paphia)*	R. Seemingly now well-established in a small area of south Cumbria.
Marsh Fritillary *(Eurodryas aurinia)*	R. Restricted to 2-3 sites in the north. Has declined and is now vulnerable.
Speckled Wood *(Parage aegeria)*	R. Occasional old records as far as the northern Lake District; now spreading in one area in extreme south, where re-introduced in 1980s.
Wall *(Lasiommata megera)*	R. Common in some coastal areas and in limestone districts of Morecambe Bay. Rarer inland, except Carlisle area.
Mountain Ringlet *(Erebia epiphron)*	R. The central Cumbria mountains are the only English localities, where it is well-established. Noted as far east as Shap Fells.
Scotch Argus *(Erebia aethiops)*	R. The two flourishing Cumbrian sites are the only English localities for this species.
Grayling *(Hipparchia semele)*	R. Rather scarce in the north, except Carlisle area, abundant in southern limestone areas; possibly declining.
Gatekeeper *(Pyronia tithonus)*	R. Quite common in west coast areas south of Whitehaven. Recently seen on Morecambe Bay coast, and occasionally inland as far as Windermere.

Meadow Brown *(Maniola jurtina)*	R. Widespread and very common.
Ringlet *(Aphantopus hyperantus)*	R. Common north of the Lake District and in west as far south as Ravenglass, also southern Eden valley; rarely elsewhere. Fig. 1, p. 17.
Small Heath *(Coenonympha pamphilus)*	R. Generally common throughout.
Large Heath *(Coenonympha tullia)*	R. Now mainly in lowland Mosses of the Solway, lower Eden valley and Morecambe Bay areas; in small numbers from more upland Pennine areas.
Monarch *(Danaus plexippus)*	IM. Six records of this American vagrant – 1980 (the only GB record that year), 1981 and 4 in 1995.

Sources and references

Beadle, H.A., (1895), Macrolepidoptera taken in Keswick and district. *Entomologists' Record* **VI**: 276-283.

Birkett, N.L., (1954), The present status of the butterfly population of the Kendal district, Westmorland. *Entomologist's Monthly Magazine* **90**: 293-298.

Birkett, N.L., (1970), Insects. Pp. 117-137 in Hervey, G.A.K. & Barnes, J.A.G., *Natural History of the Lake District.* London: F. Warne.

Dawson, G., (1879), The Butterflies of the District. *Transactions of the Cumberland Association for the Advancement of Literature & Science* **IV**: 185-188.

Day, F.H., (1901), Insects: Rhophalocera. Pp. 116-122 in *Victoria County History of Cumberland* Vol 1. Repr. 1968, London: Dawsons.

Day, F.H., (1943), The present status of Cumbrian butterflies. *North West Naturalist* **XVIII**: 284-89.

Dean, T., (1990), *The Natural History of Walney Island.* Burnley: Faust Publications.

Ellis, J.W., (1890), (revised Mansbridge, W., 1940), *The Lepidopterous fauna of Lancashire and Cheshire.* Lancashire & Cheshire Entomological Society.

Heath, J. & Emmet, A.M., (1989), *The Moths & Butterflies of Great Britain & Ireland* Vol. 7(1). Colchester: Harley Books.

Heath, J., Pollard, E. & Thomas, J., (1984), *Atlas of Butterflies in Britain & Ireland.* London: Viking.

Key, R.S. & Parsons, M.S., (1989), *Review of Invertebrate Sites in England: Cumbria. Invertebrate Site Register (Report 102).* [4 parts]. Peterborough: Nature Conservancy Council.

Kirtley, S., (undated), *The current status and ecology of the Duke of Burgundy butterfly (Hamearis lucina L.) in south Cumbria and north Lancashire.* Unpublished report for English Nature [based largely on work carried out in 1994].

Kydd, D.W., (Sub-Ed.), (1979 on), Annual insect reports in *Birds in Cumbria* (retitled to *Birds and Wildlife in Cumbria* from 1996). Cumbria Naturalists' Union (formerly Association of Natural History Societies in Cumbria.)

Lowther, R.C., (1932), Butterflies and Moths. Pp. 200-215 in Collingwood, W.G., *The Lake Counties.* London: F. Warne.

Mawson, G., (1883), The Lepidoptera of West Cumberland. *Transactions of the Cumberland Association for the Advancement of Literature & Science* **VIII**: 55-67.

Oates, M., (undated), *Survey of butterfly populations on the Carboniferous Limestone hills of the Morecambe Bay region, 1983-1985.* Unpublished report for the Nature Conservancy Council.

Routledge, G.B., (1909), The Butterflies of Cumberland. *Transactions of the Carlisle Natural History Society* **I**: 98-113.

Routledge, G.B., (1933), The Lepidoptera of Cumberland: additional species and further records. *Transactions of the Carlisle Natural History Society* **V**: 129.

Various Editors, (1956-1969), Annual reports in *The Field Naturalist (N.S.).* Penrith & District Natural History Society *et al.*

Various Editors, (1973-1978), Annual insect reports in *Natural History in Cumbria.* Association of Natural History Societies in Cumbria.

DRAGONFLIES IN CUMBRIA – a centenary review

David Clarke

Odonata, consisting of Dragonflies and their smaller relatives the Damselflies, are predatory insects at the top of invertebrate food chains. They are a small and highly accessible group, well served by identification literature for both adult and larval stages. Often requiring high quality aquatic environments, they can be valuable as indicators of both the condition and the likely conservation value of these habitats. Excluding vagrants, there are some 38 British species. 18 species are now recorded annually in Cumbria and a further 3 have occurred sporadically. The county holds nationally important populations of some scarce species.

The continuity of recording

Possibly because of the low number of species and the fact that specimens do not generally make attractive cabinet material, the group has not always attracted the attention historically accorded to some insect groups. There is nevertheless a continuous record of some value over the past century and more. Popular interest in the group has grown in recent years. Recording schemes and attractive modern publications have been important factors. The published records for the period are referenced in square brackets; asterisks (*) indicate historic specimens in the collections of Tullie House Museum, Carlisle.

Historical review

The earliest known Odonata records from the area pre-date the review period. The Carlisle naturalist T.C. Heysham (1792-1857) collected specimens and contributed notes to magazines and major entomological publishing ventures. As noted by F.H. Day [15], Heysham was aware of uncommon species such as the White-faced Darter. Specimens of the latter supplied by him from 'the north of England' to Curtis [8] were probably from the Carlisle area.

Records from the county before the turn of the century were mainly from the Lake District. Visiting Victorian entomologists included that seemingly indefatigable traveller J.J.F.X. King [22], who complained at length about the rain and did not encounter any of the area's real specialities!

The first British dragonfly book, by Lucas in 1900 [24], mentions eleven species from the counties which are now Cumbria. However, it fell to local naturalists to reveal most of the county's dragonfly interest. By the time of Lucas's book, the young Carlisle entomologist Frank Henry Day (1885-1963) had just written the first of his accounts of the Cumberland species [10]. In 1903, Day was the first to record the Keeled Skimmer in the county [11]. Another early reference to a scarce species concerns the Downy Emerald at Rydal, where it was found by Mary L. Armitt in 1904 [1].

Recording at that time tended to be on the basis of the existing county areas. During the first half of the century there were several summaries of the Cumberland species by F.H. Day [12,14,15]. His list for that county had reached 14 by 1943 and he added two more by 1947 [16,18]. (Significantly, Day is listed amongst the correspondents to the first truly 'popular' dragonfly book – by Cynthia Longfield [23]). G.B. Routledge, inveterate collector and Day's more senior Carlisle contemporary, gave a brief account of records for Westmorland and Furness up to 1933 [30]. Seemingly, and surprisingly, this remains the only review of the old county of Westmorland.

The establishment of the Freshwater Biological Association laboratories on Lake Windermere in the 1930s gave impetus to interest in aquatic insects generally. Their entomologist, T.T. Macan, was responsible for many useful Odonata discoveries in the southern Lake District and established the importance of Claife Heights as an Odonata 'hot spot', with its breeding species including the Downy Emerald [25] and White-faced Darter [21]. Ford [19] refers to this and also lists the species recorded for the Furness district up to 1953. Derek Ratcliffe [29] updated some Cumberland species information in 1955, and included the first report of what is now the most important Cumbrian colony of the White-faced Darter.

The growth of scientific and popular nature conservation, and increasing mobility, brought rapid developments in the post-war period. Biological recording nationally began in earnest after the establishment of the Biological Records Centre at Monks Wood Experimental Station, near Huntingdon. An Odonata recording scheme was in place by 1968 and, in 1977, provided the data for the distribution maps by 10 km squares in Hammond's book [20]. Carlisle naturalist W.R. Laidler (1918-1994) was one of the early contributors to the national mapping scheme.

Growth of recording activity, particularly since 1980, has substantially increased the spread of records of most species, including the rarer ones. Many of these developments have been noted in annual county reports [31]. Mapping results from the national Odonata Recording Scheme (up to 1990) are included in the latest Atlas [26].

The computerised Local Biological Records Centre at Tullie House Museum commenced in 1991. It now holds most of the available Odonata records for the county, and produces local atlases. The two maps included here have been produced from this database. A meaningful species-richness map (**fig. 1**) has only been made possible by the recent increase in recording activity. Even so, there is still insufficient data to base this on breeding status.

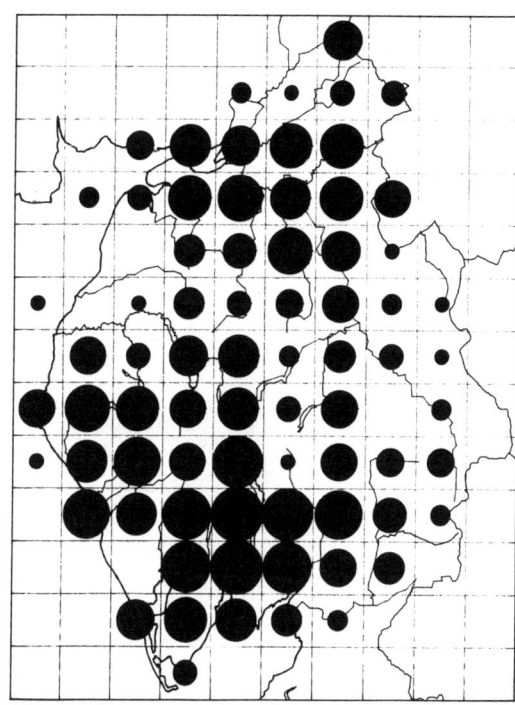

fig. 1:
Odonata: species richness post-1980

Max. 15 species per 10 km²

Changes in the Cumbrian dragonfly fauna

The absence of comprehensive data for at least the first half of the review period inevitably means that only large scale and/or very obvious changes would have been detectable: there is no proof that the county has either gained or lost any breeding species over the past 100 years. Slow trends in distribution or abundance, which would be of great interest, have gone largely undetected. Recent developments in recording noted above will hopefully rectify this for the future.

Indirect evidence, in the form of habitat losses, would suggest that many species are likely to have declined. Heathland and associated mire habitats are particularly important for Odonata. Hard figures are elusive, but a Nature Conservancy Council study [27] revealed that 19% of the area of lowland raised bogs present in Cumbria in the 1940s had been severely damaged or destroyed by the 1970s. The comparable loss figure for lowland dwarf shrub heath was 70%. These amounted to a reduction in area of some 110 square kilometres. Combined with losses at other periods, and proportionally similar reductions (of much larger areas) of upland bog and heath, the impact on dragonflies must not be under-rated.

Losses of scarce species at specific sites are referred to below; increases may also have occurred, but are less certain; the situation for the majority of species remains unproven. Migrants and casually recorded species are of interest – not least because they may prove to be portents of things to come.

Species which have suffered local extinctions:

The two Demoiselles have both undergone at least local declines. The Beautiful Demoiselle has certainly disappeared from various northern sites during this century. Day [13] knew the species from the rivers Eden and Petteril near Carlisle (NY45). There have been no records from the Eden since 1921* and the last from the Petteril is 1943*.

Day also recorded the important 'edge-of-range' populations of the Banded Demoiselle from the Solway area from 1936 onwards. The species is still present on the River Waver, where colonies apparently extend to near tidal limits. It seems to have been lost from the Eden near Carlisle (NY35)* by ca 1968. The Wigton locality (presumed R. Wampool system, NY24) is not precisely known and there have been no records from that area since the 1937 occurrence referred to by Day [15].

Losses in the two above species cannot be ascribed to any certain causes, though both are known to be sensitive to pollutants such as industrial and agricultural chemicals, sewage effluents and farm slurry, and these must remain prime suspects. The management regimes of water courses could also be a factor – certainly so in the shorter term, as one observer has noted [6].

The Variable Damselfly has apparently always been very restricted in Cumbria, seemingly preferring sites which have some neutral fen characteristics. The two known Eden valley populations, (at Newton Reigny Moss* and Cliburn Moss [specimen: Manchester Museum]), had probably both disappeared well before 1950. Newton Reigny was progressively modified by peat-cutting, a World War II drainage project and, as F.H. Day noted in 1944, enrichment caused by a Black-headed Gull colony. Derek Ratcliffe's record from Thurstonfield Lough (NY35) in 1947 perhaps represented a rare survivor of a declining population. The two current sites, in the west of the county, are relatively recent discoveries.

The nationally notable White-faced Darter is a peat-land specialist. Its mention by Lucas [24] is thought to refer to Foulshaw Moss (SD48) – a site destroyed by afforestation post-1945. Populations noted post-1950 at Oulton Moss (NY25) and Cumwhitton Moss (NY55) in the north of the county have both subsequently been lost due to drying out of pools or their occlusion with *Sphagnum*: the Cumwhitton population* persisted at least until the late 1960s [5]. A single individual seen 'over a bog in Borrowdale' – almost certainly Manesty Park (NY22), in August 1936 or 1937, may have derived from a colony never revealed and now lost (Pinniger [28] and pers. comm. 1994).

Whilst these losses are mainly attributable to human effects on land use or drainage, natural infilling of small breeding pools may also be a contributory factor. The original Claife Heights site (SD39) is now almost unsuitable, but another site found in the same area in 1993 may give longer term hope of survival [3]. The strongest colony is now that near Carlisle. Here, as often elsewhere, the breeding pools are old peat cuttings. Some management to maintain this site has recently commenced [9], and the possibility of re-introductions of the White-faced Darter at sites from which it has been lost is also under preliminary consideration.

Species which may have increased:

There is a possibility that both the Emerald Damselfly and the Common Darter may have become more widespread than they were earlier this century. Day [16,18] recorded both as 'new to Cumberland' in the 1940s, specifically mentioning his failure to find the former over the previous 40 years. That such obvious species could have been widespread and yet overlooked by him for that period seems unlikely.

The Brown Hawker was first referred to in 1973 [7] and now continues to be recorded annually on the southern fringe of the area. The first evidence of breeding was obtained in 1995.

The Southern Hawker has clearly always maintained a presence, with early records [22] pre-dating the review period. F.H. Day regarded the species as scarce. Whilst the subsequent growth of recording makes comparisons difficult, the surge of recent records (many of confirmed breeding) suggests that a real increase could be under way. This may be obscured by a variable amount of immigration in the later part of the season.

Species for which there is no evidence of significant change:

I can find no evidence to show that any of the following have undergone significant status changes, though this does not necessarily mean that no such changes have occurred:

Large Red Damselfly; Azure Damselfly; Common Blue Damselfly; Blue-tailed Damselfly; Common Hawker; Golden-ringed Dragonfly; Downy Emerald; Four-spotted Chaser; Keeled Skimmer; Black Darter.

Casual records – migrations and dispersals:

Migrations are a feature of dragonfly biology, particularly in certain species. Fluctuations in frequency of the Common Darter referred to above may be in part explained by this. Some Darters occur in Britain only as migrants, most often in exceptional summers. The Yellow-winged Darter is one of these species and single examples occurred in 1945* [17] and 1995 [31].

Movements of established British species to sites beyond their prevailing range also occur under appropriate conditions. A record of the Brown Hawker from the Loweswater area in 1954 [specimen: Liverpool Museum] is a presumed wandering of this powerful flyer. Another possible instance of medium-distance dispersal was a Downy Emerald from near Carlisle in 1967 (W.R. Laidler). The Emperor Dragonfly which was recorded at Lowick Common (SD28), and possibly other sites, in 1995 was doubtless a 'fair-weather' vagrant – its nearest regular sites being in south Lancashire.

The occurrence of the Broad-bodied Chaser at Eskmeals (SD09) is more problematic. The 'population' (max 2 adults) persisted only from 1984 to 1986 (A.B. Warburton, pers. comm.) and the possibility of accidental introduction cannot be ruled out [2]. A single 'freshly emerged' male was reported from Muncaster (SD19) in 1994 [32]. In the summer of 1960, Carlisle naturalist Ernest Blezard saw numbers of a 'powder blue Chaser' at Overwater (NY23) [D.A. Ratcliffe, pers. comm.]. Conceivably this species could have been involved – but others are possible and there have been no records before or since to resolve the question.

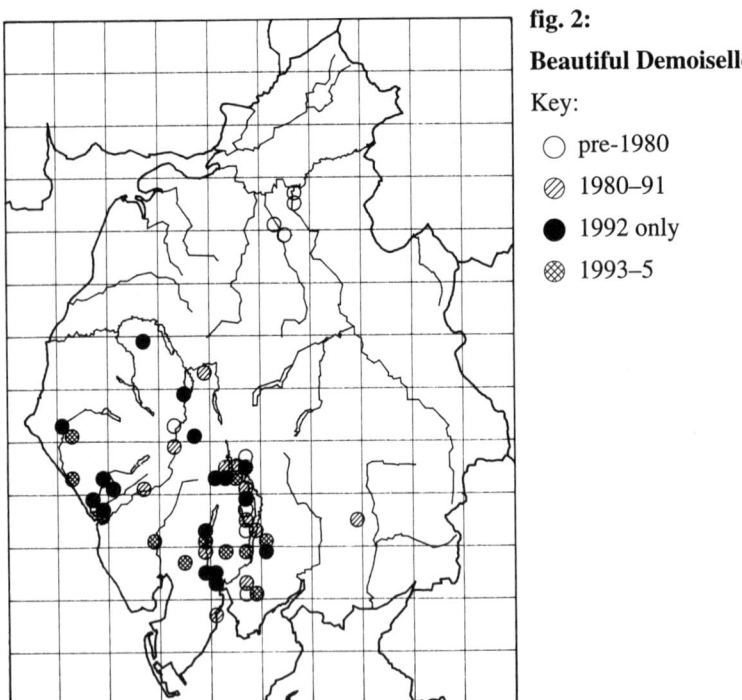

fig. 2:

Beautiful Demoiselle

Key:

○ pre-1980

◎ 1980–91

● 1992 only

⊛ 1993–5

The finding of the Beautiful Demoiselle in 5 new 10 km squares in 1992 (a 154% increase in post-1980 2 × 2 km tetrad records) may in part have reflected a local dispersal from breeding areas due to a fine early season (**fig. 2**). Similarly, the occurrence in June/July 1996 of the Banded Demoiselle in atypical habitats in the Carlisle area, as well as in apparently suitable breeding habitats on the Rivers Wampool and Eden, was suggestive of a dispersal [4]. Such events may hopefully be precursors to the colonisation of new sites, or re-colonisation of areas which have lost these species.

The future: local and national trends

There have always been pressures on dragonfly habitats, but these have greatly increased in the last few decades. Agriculture, forestry, industry, water management and leisure developments may all have contributed to the declines discussed above, and will have probably outweighed any natural trends. In addition to those which directly affect water or aquatic environments, others such as removal or modification of adjacent vegetation can prove equally significant. The existence of woodland areas close to breeding sites can be important for a few species – for example, the Downy Emerald, which is a key species in the county.

Significant and sustainable populations of some of the county's rarer species are found in variously protected situations – such as National Trust sites, Nature Reserves, SSSIs and the growing network of sites within the Lake District Environmentally Sensitive Area. This applies to the following species: Downy Emerald, Keeled Skimmer, White-faced Darter, Beautiful Demoiselle and Variable Damselfly. Whilst safeguarded to varying degrees, all such sites require careful monitoring. The species with only two known sites each, the White-faced Darter and the Variable Damselfly, must now be considered highly vulnerable. So too must the Banded Demoiselle, which is limited to a few stretches of lowland river. It is easily affected by incidents far from its immediate breeding areas, which are still incompletely known. Such species, once lost, will be most unlikely to recolonise. On a more optimistic note, there are signs that some species, nationally, are showing signs of northwards advances. This is most clearly the case with the Ruddy Darter *(Sympetrum sanguineum)*, which now occurs in Lancashire and may yet reach Cumbria. Several other species are now regularly seen well north of their ranges earlier this century.

Our understanding of the requirements and distribution of most species has increased vastly during the present century, and is now such that it possible to designate areas of high Odonata interest and to advise on appropriate management of important sites. We are now at a stage where ignorance ought no longer to be the reason why further losses are sustained. The further development of recording activities, and more specific surveys, is as essential for alerting us to these future problems as it is for the detection of natural trends.

As the century closes, predictions of global warming are being made with increasing confidence. If the rises in sea level, average temperature, cloudiness and precipitation predicted by some for northern Britain actually happen, the effects on dragonflies (as well as much other fauna and flora) in the first half of the coming century could easily exceed those of other, more direct, human influences. The most obvious trends over the past few decades have been to cold, late, springs, and only erratic occurrences of fine, settled conditions. None of these tendencies offer much for the betterment of dragonfly populations, making the care of optimal sites all the more vital.

Odonata recorded in Cumbria: current status (1995)

The present county extends over (or impinges upon) 94 10 km national grid squares. A high proportion of squares contains upland areas which support few species. By way of contrast, some of the lowlands, most especially in the south and south-west, contain an excellent variety of freshwater habitats, including mires, often with associated woodlands.

The following were designated as Key Species in Cumbria under the Odonata Recording Scheme's Key Sites Project (1988-95), reflecting the national and/or regional importance of their populations: Banded Demoiselle, Beautiful Demoiselle, Variable Damselfly, Downy Emerald, Keeled Skimmer and White-faced Darter.

The present status of all recorded species is summarised below. To give some objectivity, the number of 10k square records for each of the established species, post 1975, is categorised as follows:

Very rare	one square only
Rare	2-5 squares
Local	6-20 squares
Widespread	21+ squares

As records include casual as well as breeding occurrences, the picture presented for some species may tend, if anything, to be over-optimistic. Recording coverage is uneven, with the west and the south-east in particular requiring more in-depth work.

Beautiful Demoiselle *(Calopteryx virgo)*	Local. Mainly south-central and south-west; formerly (pre-1950) lower Eden valley.
Banded Demoiselle *(C. splendens)*	Rare. Solway plain: main sites on River Waver; formerly R. Eden (pre-1970), and possibly R. Wampool; seen 1996 on these rivers. Vulnerable.

Emerald Damselfly *(Lestes sponsa)*	Widespread, though rather less so than the other common damselfly species.
Large Red Damselfly *(Pyrrhosoma nymphula)*	Widespread and very common.
Blue-tailed Damselfly *(Ischnura elegans)*	Widespread and very common.
Common Blue Damselfly *(Enallagma cyathigerum)*	Widespread and very common. Often the only species at upland tarns.
Variable Damselfly *(Coenagrion pulchellum)*	Rare. Two sites, in W. Cumbria only; formerly at three Eden valley sites (pre-1950). Vulnerable.
Azure Damselfly *(C. puella)*	Widespread at lowland sites.
Common Hawker *(Aeshna juncea)*	Widespread, breeding in tarns at up to 2000 ft (660 m).
Brown Hawker *(A. grandis)*	Local. Breeding status uncertain; regularly seen only around Morecambe Bay from Silverdale to Barrow.
Southern Hawker *(A. cyanea)*	Local. Not recorded in the west of the county. Breeding, and possibly increasing, at lowland sites. Probably augmented by immigration in some years.
Emperor Dragonfly *(Anax imperator)*	Vagrant; at least one record in the exceptional summer of 1995.
Golden-ringed Dragonfly *(Cordulegaster boltonii)*	Widespread. Breeds mainly in upland sites in hill streams and runnels.
Downy Emerald *(Cordulia aenaea)*	Local. South-central Lake District only; many sites within this area, though breeding status little known.
Broad-bodied Chaser *(Libellula depressa)*	Vagrant and/or possible introduction; recorded at one west coast site 1984-6; single report for 1994.

Four-spotted Chaser *(L. quadrimaculata)*	Widespread. Mainly at sites with shallow, well-vegetated margins.
Keeled Skimmer *(Orthetrum coerulescens)*	Local. South-central and south-west only; few good sites known – limited to boggy seepages and vulnerable to land use changes.
Common Darter *(Sympetrum striolatum)*	Widespread. A lowland species which appears often to be augmented by immigrations.
Yellow-winged Darter *(S. flaveolum)*	Vagrant. Single records in 1945 and 1995, possibly also in 1950s.
Black Darter *(S. danae)*	Widespread. Mainly at rather acid sites, where sometimes very numerous.
White-faced Darter *(Leucorrhinia dubia)*	Rare. Two localities: one in north; the southerly one (Claife Heights) may have more than one breeding site. Lost post-c.1950 from three localities. Vulnerable.

References

1 Armitt, M.L., (1904), Neuroptera: Cordulia aenea, *etc*, at Ambleside. *Naturalist*, **1904**: 250.
2 Banks, B., (1984), *Libellula depressa* in Cumbria: a case of natural colonisation, or an accidental introduction? *Journal of the British Dragonfly Society* **1 (4)**: 47-8.
3 Clarke, D.J., (1994), The White-faced Darter Dragonfly at Claife Heights, Windermere. *Carlisle Naturalist* **2(1)**: 6.
4 Clarke, D.J. & Garner, S., (1996, in press), A dispersal of the Banded Demoiselle *(Calopteryx splendens* (Harris)*)* in the Solway Plain in 1996. *Carlisle Naturalist* **4 (2)**.
5 Cowan, C.F., (1968), *Leucorrhinia dubia* (V. d. Lind.) in Lakeland. *Entomologists' Monthly Magazine* **104**: 260.
6 Cowan, C.F., (1968), Loss of a Lakeland damselfly colony; *Agrion splendens* (Harris)) (Odonata). *Entomologists' Monthly Magazine* **104**: 262.
7 Cowan, C.F., (1973), *Aeshna grandis* (L.) (Odonata): a new record for Furness. *Entomologists' Monthly Magazine* **109**: 128.
8 Curtis, J., (1862), Leucorrhinia rubicunda. in *British Entomology* Vol. IV. London: Lovell Reeve & Co.

9 Dalglish, G., (1996), The White-faced Dragonfly management project at Scaleby Moss SSSI. *Carlisle Naturalist* **4 (1)**: 16.

10 Day, F.H., (1901), Neuroptera, Odonata, Dragonflies. Pp. 102-3 in *Victoria County History of Cumberland* Vol. 1. Repr. 1968, London: Dawsons.

11 Day, F.H., (1904), *Orthetrum coerulescens* in Cumberland. *Entomologists' Monthly Magazine* **40**: 111.

12 Day, F.H., (1917), Cumberland Dragonflies. *The Naturalist* **53**: 357-358.

13 Day, F.H., (1921), *Calopteryx virgo* L., in Cumberland. *Naturalist*, 1921: 371

14 Day, F.H., (1928), Cumberland Odonata (Dragonflies). *Transactions of the Carlisle Natural History Society* **4**: 131-134.

15 Day, F.H., (1943), Cumberland Odonata. *Entomologists' Monthly Magazine* **79**: 43-44.

16 Day, F.H., (1945), *Lestes sponsa* (Hansemann) (Odon., Lestidae) in Cumberland. *Entomologists' Monthly Magazine* **81**: 239.

17 Day, F.H., (1945), *Sympetrum flaveolum* (L.) (Odon., Libellulidae) in Cumberland. *Entomologists' Monthly Magazine* **81**: 250.

18 Day, F.H., (1947), *Sympetrum striolatum* (Charp.) (Odon., Libellulidae) in Cumberland. *Entomologists' Monthly Magazine* **83**: 289.

19 Ford, W.K., (1953), Lancashire and Cheshire Odonata (a preliminary list). *North Western Naturalist* **N.S.1**: 227-233.

20 Hammond, C.O., (1977), *The Dragonflies of Great Britain and Ireland*. London: Curwen Books.

21 Kimmins, D.E., (1943), *Leucorrhinia dubia* (V.d.L.) (Odon., Libellulidae) in the Lake District. *Entomologists' Monthly Magazine* **79**: 184.

22 King, J.J.F.X., (1882), Neuroptera of Langdale. *Entomologists' Monthly Magazine* **19**: 82-84.

23 Longfield, C., (1949), *The Dragonflies of the British Isles*. 2nd edition. London: Warne.

24 Lucas, W.J., (1900), *British Dragonflies*. London: Upcott Gill.

25 Macan, T.T., (1949), Survey of a moorland fishpond. *Journal of Animal Ecology* **18 (2)**: 160 – 186.

26 Merritt, R., Moore, N.W. & Eversham, B.C., (1996), *Atlas of the dragonflies of Britain and Ireland*. London: HMSO.

27 Nature Conservancy Council, (1987), *Research and survey in nature conservation No. 6: Changes in the Cumbrian countryside*. Peterborough: NCC.

28 Pinniger, E.B., (1937), Notes on dragonflies, 1937. *London Naturalist*, **1937**: 77-79.

29 Ratcliffe, D.A., (1955), Cumberland Dragonflies. *North Western Naturalist* **N.S.3**: 134-135.

30 Routledge, G.B., (1933), in 'The Neuroptera *etc* and Trichoptera of Cumberland, Westmorland and North Lancashire'. *Transactions of the Carlisle Natural History Society* **5**, pp. 45-6.

31 Various Editors, (1973 on), Annual insect reports in *Natural History in Cumbria* (retitled to *Birds in Cumbria* from 1977; *Birds and Wildlife in Cumbria* from 1996). Association of Natural History Societies in Cumbria (renamed Cumbria Naturalists' Union, 1990).
32 Warburton, Tony, (1994), Damsels and Dragons. *Cumbria Life* **Sept/Oct 1994**: 14-15. Carlisle: AGT.

BEETLES AND BEETLE RECORDING IN CUMBRIA

Roger Key, English Nature

Introduction

As with many other groups of animals and plants, there is an exceptional diversity of beetles living in Cumbria. The wide spectrum of habitat, temperature, rainfall regime, soil chemistry and altitude makes the Cumbrian fauna one of the most diverse in Britain. Certainly nowhere else do so many species of beetle with differing distribution patterns come so close together. In Cumbria can be found northern and montane beetles almost 'rubbing elytral humeri' with warmth-loving species of largely south-western (or even south-eastern) distribution, species typical of wet Atlantic woodland, and yet others confined to coastal dunes or saltmarsh.

Cumbria's beetle fauna – by habitat

The coast

The cliffs, dunes and saltmarshes of Cumbria have the richest coastal beetle fauna in northern Britain, with a number of species reaching their northern limit and some represented by very outlying populations.

Dunes

Cumbria's dunes support a rich fauna of typical dune beetles, such as the black strandline darkling beetles *Melanimon tibialis*, *Phylan gibbus*, and the yellowish species with dark spots *Phaleria cadaverina* – scavengers in decaying seaweed *etc*. Other scavengers include the small globular relative of the dung beetles *Aegialia arenaria* which usually burrows just below the surface of very dry sand, the strangely horned Monoceros Beetle (*Notoxus monoceros*) and the Nationally Scarce scavenger beetle *Hypocaccus rugiceps* – a small shiny black species found on dunes only on Britain's Atlantic coasts, usually in dry carrion such as dead gulls. Cumbria's dunes have a rich fauna of plant-eating species and these have been particularly well studied by John Read at Eskmeals and Ravenglass. A typical dune species is the weevil *Otiorhynchus atroapterus* – a magnificent shining black species which can occur in numbers amongst the roots of dune grasses. Another weevil, the pale grey Marram Weevil (*Philopedon plagiatus*) is very common among the fore-dunes. Some Nationally Scarce species occur on more established dunes. *Cleonus piger* is a large 'chunky' brown species on thistles; *Hypera dauci* is a much smaller but very attractively marked (and well camouflaged) species feeding on stork's-bills, and *Orobitis cyaneus* is a tiny species feeding on violets. The latter draws its legs up against its body and is superbly camouflaged as a shiny blue-black seed.

On the intertidal sand in front of Cumbria's dunes runs the tiger beetle *Cicindela hybrida*. This is a species of Britain's Red Data Book, and is restricted to a few dunes of the Lancashire and Cumbrian coast. Purple-brown in colour, this species is smaller and more elongate than the common Green Tiger Beetle of the heaths and fells. Like that species it runs rapidly over flat sand, frequently takes to the wing for short distances and is very difficult to catch or photograph!

All of Cumbria's dune systems may support equally rich faunas, but those around the Ravenglass estuary are by far the best studied.

Saltmarsh and estuary

The saltmarshes of the Cumbrian coast are relatively less rich than other coastal habitats The heavy sheep grazing that makes them so valuable for birds has reduced the vegetation to a smooth grassy lawn of fescue grass in most places, with consequently little variety of habitat for beetles. Nevertheless, one or two Nationally Scarce species, such as the small black ground beetle *Agonum nigrum,* occur in the area.

In fine grey intertidal sand in the inlets from the Solway estuary, especially on the Wampool at Kirkbride, lives a rich assemblage of ground and rove beetles. This latter site has been studied since Day's day, when the Red Data Book small burrowing ground beetle *Dyschirius angustatus* was described as new to Britain and was rediscovered there in 1992 after an absence of records of over 60 years (Key, 1993). Otherwise the species is only known from Scotland and south-east England. The Wampool seems to have the richest fauna of these burrowing ground beetles in Britain, with 6 of the 11 British species occurring there, along with 11 species of the genus of small ground beetles *Bembidion* and Nationally Scarce species such as the shiny green Sea Wormwood Weevil (*Polydrusus pulchellus*). Another peculiarity of the sands of the Solway in the past was the small ant-like beetle *Anthicus scoticus,* a Red Data Book species. This has a peculiar distribution, old records only being from Scotland and Cumbria, in contrast to all recent (post 1970) records, which are from Kent.

Cliffs

Rocky coasts usually support fewer species of beetles than sandy ones. The cliffs and coast at St Bees nevertheless have quite a rich fauna. One scarce species on the shore is the tiny water beetle *Ochthebius subinteger* (subspecies *lejolisi*) which lives in fully saline rock pools – a very unusual habitat for an insect. The cliff-top grassland at St Bees supports nesting colonies of solitary bees, which are parasitized by the ungainly blue-black oil beetles *Meloe proscarabaeus* and *Meloe violaceus*. The similar-looking, but totally unrelated, Bloody-nosed Beetle (*Timarcha tenebricosa*) feeds there on bedstraws. These species are frequently seen on the cliff top paths in spring and are, sadly, often deliberately trodden on by walkers. *Oedemera nobilis*, a metallic flower beetle with enormously enlarged hind femora in the male, reaches its northern limit here.

Limestone grasslands

The limestone grasslands to the north of Morecambe Bay around Grange-over-Sands (including Whitbarrow, Scout and Cunswick Scars and Hampsfield Fell) have a few species of beetle that have not been found elsewhere in the county. Metallic green flower beetles such as *Psilothrix viridocaeruleus* and *Oedemera lurida* reach their northern limit here. One very local species, the dark brown chafer *Hoplia philanthus*, is common on the dry grassland, also occurring on the dunes of the west coast, while on a limestone railway embankment near Church Moss is the black weevil *Zacladus geranii* which feeds on Bloody Crane's-bill. There are a number of other plant-eating species on these calcareous grasslands, particularly spectacular ones being the bright metallic green leaf beetles *Cryptocephalus aureolus* and *Cryptocephalus hypochaeridis*, which feed on small yellow composites such as *Hypochaeris radicata*. The related black and rich-yellow species *C. moraei* feeds on St John's-worts. An enigma of these coasts and grasslands is the absence of the bright yellow Sulphur Beetle (*Cteniopus sulphureus*) which has, to my knowledge, never been found in Cumbria. A suite of species with which it is commonly found in its southerly range all reach their northern limit on Cumbria's dunes and calcareous grasslands, so perhaps one or two colonies may be there to discover in the future.

Wetlands

The beetles of the remains of the raised mires around the Solway and Morecambe Bay have been less studied than those in other similar habitats such as in the Humberhead Levels or Wales, although the Nature Conservancy Council carried out a general entomological survey of some Cumbrian mires in 1988. The mossland itself is a naturally species-poor habitat, but a few species of beetle are specific to the highly acidic conditions of *Sphagnum* bogs. Two highly colourful ground beetles, the spectacular *Carabus nitens* – quite large and shining gold-green with longitudinal ridges, and the smaller Nationally Scarce *Agonum ericeti* – golden green with metallic red elytra, are found in the Cumbrian mosses, along with the less spectacular but also scarce *Cymindis vaporariorum*. A number of scarce water beetles are also typical of the highly acidic conditions, such as *Agabus unguicularis* and *Hydroporus obscurus*.

Slightly less acidic wetlands, including the edges of the mosslands and the edges of lakes and tarns, usually have richer faunas, not least because of the increased diversity of vegetation at such sites. While often a sign of deterioration of habitats, invading birch and sallow scrub also supports scarce species such as the handsome metallic blue leaf beetle *Cryptocephalus parvulus*.

Cumbria has one of the richest county faunas in Britain of the spectacular metallic multicoloured reed beetles of the subfamily *Donaciinae*, whose larvae form air-filled root galls on various aquatic plants. There are at least eight species of *Donacia* and three species of *Plateumaris*. In addition, the fully aquatic species *Macroplea appendiculata* has been found only at Talkin Tarn in recent years, although there is an old record from the River Eden. *Donacia aquatica* used to be widespread in Britain, but its only Cumbrian localities at Tarn Howes and Braithwaite Moss are now the last colonies south of the Scottish border.

That much still remains to be discovered in Cumbria's wetlands was exemplified in 1993 by the discovery of the waterpenny beetle *Psephenus palustris* in a small flush in Smardale Gill, representing a whole family of beetles (*Psephenidae*) new to Northern England.

Cumbria's water beetles have been better studied in latter years than much of the rest of its beetle fauna. Talkin Tarn has become a 'classic' locality for the study of water beetle biogeography and ecological genetics. Studies by Shirt (1981) have revealed much about the post-glacial colonization of north-west Britain by water beetles. Cumbria is one of the few areas in England for the whirligig beetle *Gyrinus natator*, most records of which refer to the closely related species *G. substriatus*, which used to go under that name. Other important water beetle species include the Red Data Book riffle beetle *Stenelmis canaliculata*, which has been found among stones on the shores of Windermere and is otherwise only found at two sites in East Anglia. Two small Red Data Book species typically only found in undisturbed relict fen and bog are *Hydroporus scalesianus* at Biglands Bog and *H. rufifrons* from Thurstonfield Lough and the Windermere south fen. The high tarns and streams have a few truly montane water beetles such as *Agabus arcticus*, *Agabus congener* and *Hygrotus quinquelineatus*.

Rivers and streams

Cumbria's rivers support a diversity of water beetles dependent particularly on the stoniness of the river bottom and the swiftness of current. Typical species include various riffle beetles (*Elminthidae*) – Cumbria having one of the highest diversities of this family anywhere in the country. The yellow and dark-brown spotted *Platambus maculatus* and the unusual Hairy Whirligig (*Orectochilus villosus*) are also typical. The latter gyrates at night on the surface of still pools in rivers and leaves the water as a larva to pupate halfway up a tree.

The shoals of shingle, sand and silt at the margins of Cumbria's rivers support a diverse fauna of specialized ground and rove beetles. This fauna used to be widespread in Britain but has disappeared from much of the country through pollution and river engineering, such that only the middle reaches of rivers in the north and west of Britain still support good faunas. Two small species of ground beetle, *Bembidion schueppeli* and *Bembidion stomoides*, were added to the British list by Bold & Murray, who found them on the banks of the River Irthing at Lanercost Abbey in the mid nineteenth century. The former has an unusual distribution, occurring only on fine silts of rivers near either side of the Scottish border. The Irthing and Eden have been shown to have particularly good faunas, with a high diversity of ground beetles in the genus *Bembidion*, as well as the minute click beetles *Zorochros minimus* and *Fleutiauxellus maritimus* and several species of the bug-eyed rove beetles of the genus *Stenus*. In mossy waterfalls there is an almost certainty of finding the related species *Dianous coerulescens* – similar in appearence but metallic blue with an orange spot on each elytron. There are some excellent lists from Cumberland's rivers in Day's monograph and there is almost certainly much to be discovered, and rediscovered, on Cumbria's river margins.

The woodlands

Cumbria's damp Atlantic oak-woods are home to a number of scarce species, the most important of which is the brown weevil *Procas granulicollis*, which feeds on Climbing Corydalis (*Ceratocapnos claviculata*), and which has never been found outside the British Isles. A Red Data Book species, it has been found on the Cardiganshire and Northumberland coasts, as well as near Bassenthwaite in Cumbria by John Read (initially identified as *P. armillatus* (Read, 1989)). John Read also discovered another weevil, the dull red *Furcipus rectirostris*, new to Britain, near Hallbolton Bridge near Gosforth in 1979. This species feeds on developing stones in the fruits of Bird Cherry (*Prunus padus*). It has subsequently been found in Yorkshire, Scotland and Wales. It is unclear whether it has been overlooked in the past or is a recent colonist. Another speciality of woodland in Cumbria is the large bronze tree-climbing ground beetle *Calosoma inquisitor*, which preys particularly on caterpillars of the green Oak Tortrix Moth, and is otherwise only found regularly in central Wales and the New Forest. The large deep-black ground beetle *Pterostichus cristatus* has a peculiar distribution in Britain, being restricted to woodland in a band across northern England from Co. Durham to east Cumbria.

Cumbria's dead wood beetle fauna is not exceptionally rich, although some species such as the colourful Black-headed Cardinal Beetle (*Pyrochroa coccinea*) are commoner in Cumbria than much of the rest of England. The Red Data Book false click beetle *Dirrhagus pygmaeus* – a small brown species, is found in woods in south Cumbria, as is the bizarre-looking wood-boring weevil *Trachodes hispidus* – covered with light and dark erect scales, which reaches its northern limit here. One quite spectacular wood-boring weevil is *Mesites tardyii* – a large shiny chestnut black species. This tends to bore into very solid timber and is found mainly in coastal woodland such as the Muncaster woodlands on the west coast, where it is common. It has been speculated (J. Marshall, quoted in Read 1982) that the species is dispersed in marine driftwood.

Another very scarce species in Cumbria's woodlands is the dung beetle *Aphodius nemoralis*, which specializes in deer dung in shady woodland and has been found in the woods around Coniston. Otherwise it is primarily a Scottish species.

The fells and mountains

Cumbria's mountains are sufficiently high and northerly to have a few truly boreo-montane species in their fauna. Among these is a black ground beetle *Nebria nivalis*, a species which frequents the edges of late-lying snow and, with a form of 'anti-freeze' in its body fluids, ventures out onto the snow to predate less hardy species immobilized by the cold. It is only recorded in England from the summit of our highest mountain, Scafell Pike (Key, 1981), and recently from Cross Fell in the north Pennines (Downie *et al.*, 1994). Other montane ground beetles include *Leistus montanus* – a species recorded from Skiddaw, Langdale Pike and at surprisingly low altitude at Wan Fell, and *Elaphrus lapponicus* which occurs on seepages through moss on Skiddaw. The rove beetle *Olophrum assimile* is only known in Britain from Cross Fell and a single site in Scotland (Hyman & Parsons 1994). These are all Nationally Scarce or Red Data Book species and, with the exception of *Nebria nivalis*, which occurs on Snowdon, reach their southern limit in Cumbria. Two montane weevils, *Otiorhynchus arcticus* and *O. nodosus*, have also been recorded from Cumbria. *O. arcticus* is only known from Wasdale Screes in the county (Read, 1991), while *O. nodosus* is rather more widely spread among the fells.

There are many other less scarce northern species on the tops. Under stones and among *Racomitrium* moss on mountain summits, as well as at the edges of highland streams, the ground beetle *Nebria gyllenhali* is very common and is very similar in appearance to *N. nivalis*. Other northern ground beetles are *Patrobus assimilis*, *Pterostichus adstrictus*, the pill beetles *Byrrhus fasciatus* and *B. arietinus* and the rove beetles *Geodromicus longipes* and *Anthophagus alpinus*. The two pill beetles much resemble sheep droppings and are the only beetles to eat the *Racomitrium* itself. Less common are the burrowing ground beetle *Miscodera arctica* – with front legs like a Mole's, and other montane ground beetles *Notiophilus aesthuans* (among small stones) and *Patrobus septentrionalis* (along upland streams).

On the moorlands, as opposed to the summits, there are many other northern species to be found. Among sheep droppings the black and maroon dung beetle *Aphodius lapponum* is common, along with several other species also found lower down and further south. The common metallic blue- or violet-black 'Dumbledor' on the moors is usually *Geotrupes stercorosus*, a typical northern and western species. Occasionally the much scarcer very shiny *Geotrupes vernalis* can be found in abundance, especially in the spring, along with the spectacular Minotaur Beetle (*Typhaeus typhoeus*), which has three very prominent horns on the thorax of the male.

Another very conspicuous species is the click beetle *Ctenicera cuprea*, which can be either entirely metallic purple or have dull orange elytra. I have seen these beetles, along with the metallic green and brown Bracken-clock (*Phyllopertha horticolor*) at densities of around 20 per square metre in June in upper Wasdale throughout huge areas. Being root feeders, they are regarded as destructive species, lowering the productivity of the sheepwalk, but they are very important food for upland birds. Four other conspicuous moorland species are the very large ground beetles of the genus *Carabus*. These come out in numbers at night but also in daytime during and after a summer rain which breaks a dry spell. They can also reach high densities, sometimes of one to every few square metres, as I once observed on the moors below Stickle Pike. Most common are the two violet ground beetles *Carabus problematicus* and *C. violaceus*, although the coppery-coloured *C. arvensis* is almost as common. Much less frequently seen is the dull blue-black *C. glabratus*, which usually turns up only as occasional specimens.

Beetle recording in Cumbria

Historically, recording of Cumbria's fauna has been very extensive, although patchy, with by far the greatest amount of recording during last century and the early part of this being in the northern part of Cumberland. Carlisle and its Natural History Society was an important centre of this activity. Southern Cumberland, most of the Lake District, and all of the Westmorland and Lancashire (vc. 69) parts of modern Cumbria were relatively poorly studied in comparison, and there are relatively fewer early data from these southern areas to form a baseline from which to study trends in distribution and abundance.

The early history of beetle recording in Cumberland is outlined by Day (1909) in the introduction to his monograph of the county's beetles (Day, 1909 – 1933). He relates that the earliest coleopterists working in the county were the indigenous Thomas Heysham (1791-1857) of Carlisle, Mr Weaver, James Dale (1792-1872) from Dorset, Dr Z.M. Leach and the Durham coleopterist Thomas Bold (1816-1874).

As was common in Victorian times, the locality data left by some of these early coleopterists are sparse, often referring only to 'Solway' or 'Carlisle District', which could mean anywhere within a thirty mile radius, or the name of the parish in which the records were made. Fortunately Bold's and Heysham's data are rather more specific and they left a very good legacy of information on the early and mid-Victorian beetle fauna of northern and eastern Cumberland. All of these entomologists communicated widely with the national experts. Cumberland beetle data are extensively quoted in the early major works on British insects and

British beetles such as Curtis (1824-1838), Stephens (1827-1835), Dawson (1854) and Fowler (1887-1891).

In the latter part of the nineteenth and early part of the twentieth century, another generation of coleopterists, this time largely indigenous to (or at least living in) Cumberland, carried out a prodigious amount of recording in the county. These were George Routledge (1864-1934) from Tarn Lodge, near Carlisle, James Murray (1872-1942), Harry Britten (1870-1954), and Frank Henry Day (1875-1963) from Carlisle. Murray was born in Carlisle but lived at Kelswick near Wigton before retiring across the border to Gretna. Britten was born in Wiltshire but lived and worked in Cumberland until 1913, and retained links with the county while he worked at Manchester Museum until 1937. Day produced the most important works on Cumberland beetles to date; the list in the Victoria County History and the serial monograph in this journal (Day 1909-1933). Britten's Cumberland material resides in Manchester Museum and Day's, Routledge's and Murray's collections are in Tullie House Museum in Carlisle.

Subsequent workers, mainly in north Cumbria, have been William Davidson from Penrith (working in the 1960s, his collection is now at Liverpool Museum), John Read of Whitehaven (concentrating particularly on the weevils and leaf-beetles) and David Atty, who moved to Cockermouth in 1988 after documenting the beetle fauna of Gloucestershire.

Early beetle recording in Westmorland was also documented by Day (1918). Earlier papers with Westmorland records were sometimes published in *The Naturalist* – the natural history journal for the North of England produced by the Yorkshire Naturalists' Union. Frank Day and Harry Britten's records dominate this publication, but there are many records from Mr T. Bowman of Tebay. Other records quoted are largely from visitors to the county – the Scotsman James E. Black, whose collection is now in the Royal Scottish Museum, the Reverend T. Blackburn of Islington and Mr M.L. Thompson. There was no Victoria County History of Westmorland. In the 1930s the late T.T. Macan of the Freshwater Biological Association (now the Institute of Freshwater Ecology) carried out extensive studies of the water beetles in the Lake District (Macan, 1940), while in more recent years Neville Birkett and Tim Bird have recorded Coleoptera in this part of Cumbria as a part of wider studies on the insect fauna.

That part of Cumbria formerly in Lancashire is much more difficult to investigate and the *Victoria County History* for Lancashire, although having a section on Coleoptera, gives virtually no information from north of the Lune. There are some records scattered through the *Transactions of the Lancashire and Cheshire Entomological Society* which has, however, similarly concentrated on the more southerly parts of Lancashire.

Being so rich for beetles, and also being a popular holiday destination, Cumbria has often been visited by coleopterists from other parts of the country. The results of their work are widely scattered in their collections and notebooks, only occasionally being published.

There have been a number of specific studies and surveys of Cumbrian Coleoptera in relatively recent years. David Shirt (1981) has studied the evolutionary relationships among water beetles in various Cumbrian lakes. The National Trust's biological survey team undertook extensive surveys of all aspects of the wildlife of most Trust properties in Cumbria in 1980 and subsequent years. They produced extensive series of reports on their findings and include many records of Cumbrian beetles by Keith Alexander and David Shirt.

In 1989, the Nature Conservancy Council's *Invertebrate Site Register* (Ball, 1994) collated information on scarcer species of invertebrates, including 660 Red Data Book, Nationally Scarce and Local species of beetle, and on rich sites throughout the county (Key & Parsons, 1989). Red Data Book and Nationally Scarce species of beetle are listed, and the concepts behind these statuses defined in Hyman & Parsons (1992 & 1994).

In June 1987 the informal Coleoptera Recording Group held its annual recording meeting at the Rowrah Field Centre in the west of the county (Angus, 1987) and 19 coleopterists and five other entomologists from all over Britain spent two-three days (unfortunately largely in pouring rain!) recording beetles in this under-recorded part of the county. This resulted in the discovery of one species of beetle, a rove beetle *Hadrognathus longipalpus* (with extraordinary huge sickle-shaped mandibles), new to Britain (Lott, 1989) – although it has subsequently turned up in Wales.

A joint meeting of that group with the British Entomological and Natural History Society similarly held a long weekend meeting at Castle Head at Grange-over-Sands in June 1993 and was attended by 15 coleopterists and four other entomologists. The results of both of these meetings are being collated into a report to be produced by English Nature (Key and Drake, in prep.).

Conservation and Cumbria's beetles

There are many species of beetle that have not been seen in Cumbria for many years and may well be extinct in the county. These include the attractive red and black fungus-dwelling darkling beetle *Diaperis boleti* which was found at Dalston Hall Wood by Heysham and has not been seen since in Cumbria (and only a few times elsewhere). The click beetle *Cardiophorus graminis* and the attractive wasp-like longhorn beetle *Plagionotus arcuatus* were both known from Baron Wood. All of these species are now very rare nationally, and the two latter are probably now extinct. We will never know if the leaf beetle *Phyllodecta polaris* ever occurred in Cumbria. Its foodplant, the Dwarf Willow (*Salix herbacea*), was already scarce in Cumbria through overgrazing by sheep when the beetle was first discovered to be present in Britain in the Scottish Highlands.

Cumbria's beetle fauna remains very rich, and, with so much of its land area under conservation ownership or fairly traditional management, it is unlikely that very many other species have genuinely become extinct in the county. Changes in land use, including the recently much-increased levels of sheep grazing on the fells, drainage and/or afforestation of moorland and interference with the natural fluctuations in flow-rate of Cumbria's rivers, are the human activities most obviously likely to endanger Cumbria's beetle fauna. Longer term trends, in particular climatic change induced by the burning of fossil fuels, may eventually favour some southern species which currently inhabit only the coast and warmer areas (*eg* around Witherslack). These changes might be at the expense of rendering the uplands unsuitable for the truly boreal species. Cumbria, with its mixture of 'northern', 'southern' and 'western' species, is exceptionally placed for monitoring the effects of climate change, if the predictions of global warming become a reality.

Acknowledgements

I would like to thank John Read for supplying information on his records towards the preparation of this paper.

References

Angus, R.B., (1964), Some Coleoptera from Cumberland, Westmorland and the northern part of Lancashire. *Entomologist's Monthly Magazine* **100**: 61-69.

Angus, R.B., (1965), Further Coleoptera from Cumberland, Westmorland and the northern part of Lancashire. *Entomologist's Monthly Magazine* **101**: 4-8.

Angus, R.B., (1987), Coleopterists Field Meeting, West Cumbria, 26th-28th June 1987. *Antenna* **11**: 151-152.

Ball, S.G., (1994), The Invertebrate Site Register - objectives and achievements. *British Journal of Entomology and Natural History* **7**, Supplement 1: 2-14.

Curtis, J., (1824-1838), *British Entomology*. 16 vols. London: Lovell Reeve & Co.

Dawson, J.F., (1854), *Geodephaga Brittanica. (A Monograph of the Carnivorous Ground Beetles Indigenous to the British Isles)*. London.

Day, F.H., (1909), The Coleoptera of Cumberland, Part 1. *Transactions of the Carlisle Natural History Society* **1**: 22-150.

Day, F.H., (1912), The Coleoptera of Cumberland, Part 2. *Transactions of the Carlisle Natural History Society* **2**: 201-256.

Day, F.H., (1918), Westmorland Coleoptera. *The Naturalist* **43**: 189-191, 224-226, 285-288, 389- 391.

Day, F.H., (1919), Westmorland Coleoptera. *The Naturalist* **44**: 77-79, 239-242, 327-328.

Day, F.H., (1923), The Coleoptera of Cumberland, Part 3. *Transactions of the Carlisle Natural History Society* **3**: 70-106.

Day, F.H., (1928), The Coleoptera of Cumberland, Part 4. *Transactions of the Carlisle Natural History Society* **4**: 135-136.

Day, F.H., (1933), The Coleoptera of Cumberland, Part 5. *Transactions of the Carlisle Natural History Society* **5**: 117-125.

Downie, I.S. *et al.*, (1994), The Invertebrates of Cross Fell and Dun Fell summits, Cumbria. *The Vasculum* **79 (3)**: 48-62.

Fowler, W.W., (1887, 1888, 1889,1891), *The Coleoptera of the British Islands*. 5 vols. London: Reeve.

Hyman, P. & Parsons, M., (1992), *A review of the scarce and threatened Coleoptera of Great Britain, Part 1. UK Nature Conservation No. 3.* Peterborough: JNCC.

Hyman, P. & Parsons, M., (1994), *A review of the scarce and threatened Coleoptera of Great Britain, Part 2. UK Nature Conservation No. 12.* Peterborough: JNCC.

Key, R.S., (1981), *Nebria nivalis* (Payk.), (Col.: Carabidae) on Scafell Pike, Cumbria. *Entomologists Monthly Magazine* **116**: 160.

Key, R.S., (1993), *Dyschirius angustatus* (Ahrens) (Carabidae) and other Coleoptera from the Wampool Estuary, Kirkbride, Cumbria. *The Coleopterist* **2**: 29-30.

Key, R.S. & Drake, C.M., (in prep.), *The results of three entomological recording meetings in Cumbria between 1987 and 1993*. English Nature Research Reports.

Key, R.S. & Parsons, M.S., (1989), *Review of Invertebrate Sites in England: Cumbria. Invertebrate Site Register (Report 102)*. [4 parts]. Peterborough: Nature Conservancy Council.

Lott, D.A., (1989), *Hadrognathus longipalpis* (Mulsant & Rey) (Coleoptera : Staphylinidae) new to the British Isles. *Entomologist's Gazette* **40**: 221-222.

Macan, T.T., (1940), Dytiscidae and Haliplidae (Coleoptera) in the Lake District. *Transactions of the Society for British Entomology* **7**: 1-20.

Penny, S. & Key, R.S., (1994), 'Entscape': Computerized Invertebrate bibliography. Section Invertebrates 8.3. in Gent, A., *Species Conservation Handbook*. Peterborough: English Nature.

Read, R.W.J., (1981), *Furcipus rectirostris* (L.) (Coleoptera: Curculionidae) new to Britain. *Entomologist's Gazette* **32**: 51-58.

Read, R.W.J., (1982), *Mesites tardii* (Curtis) (Coleoptera: Curculionidae) new to West Cumbria, with notes on the species in Britain. *Entomologist's Gazette* **33**: 233-242.

Read, R.W.J., (1989), *Procas armillatus* (F.) (Curculionidae) from Cumbria. *The Coleopterist's Newsletter* **35**: 7.

Read, R.W.J., (1992), *Otiorhynchus arcticus* (Fab.) (Col.: Curculionidae) in Cumbria. *Entomologist's Record* **104**: 80-81.

Shirt, D.B., (1981), *Potamonectes depressus-elegans* in the British Isles. *Balfour-Browne Club Newsletter* **21**: 2-6.

Stephens, J.F., (1827-1835), *Illustrations of British Entomology. Mandibulata.* Vols 1-5. London.

BIRDS OF PREY IN CUMBRIA

Geoffrey Horne

Cumbria has always been the stronghold for predatory birds in England. The high precipitous crags have long been home for Peregrines and Ravens, and in this century have seen the return of the Golden Eagle. The lower crags are a favourite haunt of Buzzards, whilst the few remaining heather moors support Merlins, and sometimes Hen Harriers. Woodlands, including the much-maligned conifer plantations, are habitats for Sparrowhawks and have recently gained the much rarer Goshawk. Kestrels range widely over the hills and lowlands alike.

The writer has been closely involved in monitoring many of these species over the past 30 years and more. This work has been much drawn upon in the species accounts below. As an important member of the diurnal predator fauna, the Raven has been part of these studies, and is thus included as an 'honorary' bird of prey.

Species which have occurred only as non-breeders are not considered here.

Historical overview

The first written records of birds of prey in the county appear in the medieval period, when rights to certain species for falconry were jealously guarded, and breeding sites thus enjoyed a certain measure of protection. A document of the 1270s refers to '... *aeris aquilarum, accipitr', nisor',* ...' in Denton (N.E. Cumbria) and is highly suggestive of the breeding presence of Golden Eagles, Goshawks and Sparrowhawks (although such identifications must always remain somewhat conjectural). Much more recognisable specifically are *'falcones'*, mentioned in several documents of the Priory of St Bees. From these it is clear that Peregrines were present on St Bees Head in the late 14th century; a mid-16th century document from the same source includes mention of these birds being presented to the sovereign.

With post-medieval changes in upland management – especially for sheep husbandry and for game preservation – predatory birds were increasingly viewed as competitors. The story of their fate, especially over the last two centuries, is well understood, and the Cumbrian situation is much as elsewhere in upland Britain. The most spectacular raptors – the White-tailed and Golden Eagles – had probably both gone by the end of the eighteenth century; Ospreys also once bred, though their demise is less well charted; the Red Kite and Marsh and Hen Harriers were probably the last to disappear as breeding species – though still well before the present review period. Macpherson's great survey at the end of the last century – *'A Vertebrate Fauna of Lakeland*' (Macpherson, 1892) – is the main source for details of this dark period of the county's history, and is referenced elsewhere in this account by the author's name or (M).

In the later phases of this onslaught, the influences of the Victorian vogue for collecting played a significant part. Eggs of most species were systematically removed from traditional nest sites, often annually; adults and young were shot, destined as subjects for the developing art of the taxidermist. A largely unplanned benefit of this wholesale plunder has been the eventual transfer of some of this 'time-sample' into the permanent care of public museums, where its potential for research and education is valued; the Cumbrian material held at Tullie House Museum, Carlisle, is a notable instance.

The tail-end of these depredations persisted well into the present century, and must have depressed the populations of most species. An early germ of the growing pressures for conservation was the formation, by members of the Carlisle Natural History Society, of the Cumberland Nature Reserve Association. The protection of wildlife outwith nature reserves was one of its aims, and in 1914 a payment-by-results scheme to warden nest sites of Ravens, Buzzards and Peregrine Falcons against egg-collectors was started (CNRA, 1915).

The two wartime interludes reduced keepering pressures, but the rise of the conservation movement and successive developments in legislation post-1950 have provided the real basis of improvements. Greater tolerance and protection in the second half of the century encouraged the return of three species long absent: the Golden Eagle re-colonised at the height of the toxic chemicals era – and survived; aided by re-introductions, the Goshawk began to spread in the Border forests from the 1960s onwards; at much the same time, the Hen Harrier returned naturally to some remoter parts of the county.

Conversions of heather moor to grassland, and recent conifer afforestation of large areas of rough pasture and bog have greatly changed many of the uplands – with significant permanent effects on their ecology. Loss of heather moor has reduced the potential habitat for the specialised Merlin; the hunting grounds of several other raptors have also been much diminished. Although the longer-term effects on moorland wildlife generally are adverse, the early (scrub) stages of the new plantations have provided temporary vole-rich feeding grounds and encouraged some species.

A situation of urgency developed in the 1950s and 60s with the appearance of a new and insidious threat – toxic chemicals. Pesticides, introduced into food chains via seed dressings and sheep dips, had a severe impact on many predators. The main chemical residues involved were DDE (the main breakdown product of DDT) and HEOD (the derivative of aldrin and dieldrin). Accumulated doses via prey species built up sub-lethal concentrations in the liver and other organs. This affected the central nervous system of the birds, resulting particularly in increased adult mortality (from the dieldrin group) and eggshell thinning (from

DDT). Observations on Cumbrian raptors, especially Peregrines, contributed to the painstaking process of establishing the linkage between the population decline, eggshell thinning and organochlorine pesticides – see especially Ratcliffe (1970).

The populations of Peregrine Falcons, Merlins and Sparrowhawks suffered much in this episode. But for the voluntary restrictions on usage of pesticides which were introduced in 1962, 1964 and 1966, these species might even have been lost as breeders in the county. Toxic residues affected other species to varying extents, and monitoring of their levels continues to the present day. The recovery of the affected species proceeded through the 1970s and 1980s, spectacularly so in the case of the Peregrine. The latter has now achieved unprecedented population levels nationally, and most notably in Cumbria.

Reduction of the influences of toxic chemicals has at least diluted the effects of other pressures. Despite legislation, intolerance from sporting interests remains in some parts of the county. The species which has been most critically affected is the returning Hen Harrier. The taking of eggs, and in some instances young, for collections or falconry has also continued. The post-war growth of outdoor recreation has greatly increased the disturbance to breeding birds. Fell walking and rock climbing have been the main concerns, but over-zealous bird-watching may itself have contributed. Wardening of key sites during the nesting season has had to cope with these pressures, and especially the more overtly felonious activities. In addition to some notable efforts by volunteers, the Royal Society for the Protection of Birds has been heavily involved, not least in the protection of the Haweswater eagle site from 1970 onwards. The growth of popular interest in birds, especially since 1970, has ensured that changes in the fortunes of birds of prey are now much more closely monitored. The formation of the Cumbria Bird Club and its Raptor Study Group in 1992 were important milestones in these developments.

Surveys by the British Trust for Ornithology in 1968-72 and 1988-91 also absorbed much local enthusiasm, and led to publication of breeding atlases (Sharrock, 1976; Gibbons *et al.,* 1993). These gave a valuable national perspective to the importance of Cumbria as a haven for British birds of prey.

Species which last bred before 1900

White-tailed Eagle *(Haliaeetus albicilla)*
Macpherson records that White-tailed Eagles bred at Wallow Crag (Haweswater), Eagle Crag (Borrowdale), Eagle Crag (Buttermere), Buck Crag (Martindale) and other sites up to the late 18th century. There are several Erne, Iron and Heron Crags in the Lake District, and such names have been linked

primarily with this species. However, in early accounts it is often difficult to separate the White-tailed from the Golden Eagle, which was also present, but has a less clearly defined history.

Osprey *(Pandion haliaetus)*

In the late 18th century, Francis Willughby stated that *'the Sea Eagle or Osprey, Haliaetus sive Ossifraga which preys often upon our rivers; there is an aery of them in Whinfield Park, Westmorland, preserved carefully by the Countess of Pembroke'* (M). The area referred to is near Penrith. Ullswater was another probable breeding site referred to by Heysham (1794). The Osprey was known only as a passage visitor by the time of Macpherson, and so it has remained.

Marsh Harrier *(Circus aeruginosus)*

It is clear that Heysham (*op. cit.*) knew the Marsh Harrier [as the Moor Buzzard] from personal experience, stating that *'this bird is very frequent upon our moors'*. Macpherson considered this species probably survived up to the mid-19th century in remoter parts of the county. He quotes a Mr Hutchinson who believed the species once bred regularly on Hay Fell [near Kendal].

Red Kite *(Milvus milvus)*

The Red Kite was referred to as breeding in the woods near Armathwaite, by Dr Heysham in 1796 (M). Ullswater, Derwentwater and Windermere are all quoted as strongholds of this species in the 18th century. A specimen from Portinscale, Keswick, shot in 1840, is now in Tullie House Museum. Macpherson considered it to be *'probably the last of the indigenous Kites that inhabited the Lake District from prehistoric times'*.

Species accounts: rare breeders in the 20th century

Three species whose main British range is well south of Cumbria have made very rare attempts to breed during the review period: Honey Buzzard, Montagu's Harrier and Hobby.

Honey Buzzard *(Pernis apivorus)*

Apart from a tenuous record of breeding at Lowther in the eighteenth century via Dr Heysham (M), there is no evidence of the former breeding of this species. Possible breeding presence was reported by Blezard (1943) from a north Cumberland wood in 1913 and again in 1917. Such occurrences had evidently ceased by 1925. There have been no further reports of breeding attempts, and only rare sightings of single birds since then.

Montagu's Harrier *(Circus pygargus)*

Macpherson considered this species always to have been the rarest of the harriers in Lakeland, and gave no evidence of it breeding. A pair in north Cumberland in 1923 might have bred but for loss of the male, 'killed accidentally' (Blezard, 1943). (Breeding presence of this species elsewhere in northern England in the 1950s was too short-lived to affect Cumbria).

Hobby *(Falco subbuteo)*

Macpherson had only two reports (of single birds) from Cumberland, and none from the remainder of the district. Blezard (1943) records the claim of a farmer that a pair had nested at Hutton [near Penrith]. There were several possible breeding attempts in the 1930s in north Cumberland. The species still occurs annually, if sparingly, on passage.

Species accounts: regular breeders

The following species have all bred annually for at least part of the century, and are given more detailed treatment below: Golden Eagle, Buzzard, Sparrowhawk, Goshawk, Hen Harrier, Peregrine Falcon, Merlin, Kestrel and Raven.

Golden Eagle *(Aquila chrysaetos)*

Macpherson stated that Golden Eagles, as well as White-tailed Eagles, had bred in Lakeland before the end of the eighteenth century. He listed several former Lake District eyries, as well as references to the Bewcastle Fells. From the beginning of this century, there were increasing reports of birds being seen throughout the county. Doubtless, most related to unattached or juvenile birds looking for a mate and a suitable breeding territory. Blezard (1958) correlated this increase to *'the re-occupation of not so distant breeding haunts'* [in S.W. Scotland].

A pair of birds had settled into a territory in the Haweswater area by 1957. Here, on 23 March 1957, Derek Ratcliffe saw a Golden Eagle leave a crag, on which he discovered a newly-built nest. A visit two weeks later revealed an empty nest and no Eagles, but a single bird was seen about the crag on a further visit in April (Blezard, 1958). Elsewhere in the county, a second unused nest was found in late 1957 by R.J. Birkett on a big crag in a western valley. Although this nest was repaired annually each spring, no eggs were ever seen and these birds had disappeared by 1965.

Before 1965 any Eagles in the area were almost certainly affected by toxic chemicals used in sheep dips (via sheep carrion prey). During these difficult years, the Haweswater birds continued to hold their territory and each year made up a number of nests, but there are no records of eggs being laid. Following the withdrawal of dieldrin sheep dips in 1966, the fortunes of Cumbrian Golden Eagles improved dramatically. On 29 March 1969 the first recorded egg in England for over two hundred years was found in the original nest. A second egg was laid on the 4th April. Unfortunately this nest site was close to two very popular footpaths and unintentional disturbance over the Easter weekend may have been the cause of desertion. The eggs were removed under a Nature Conservancy Council licence, for analysis by the Institute of Terrestrial Ecology. They were found to be fertile and contained only low levels of toxic chemical residues, and are now in the collection of Tullie House Museum. As a result of this failure and the sensitivity of these birds, a 24-hour protective guard was mounted on the site. No one was allowed near any of the usable alternative nests during the breeding season.

The following year (1970) the birds moved into an adjacent valley and made up an alternative nest. Eggs were laid during the last week in March and after six weeks of incubation a single chick was hatched during the first week of May. The parent birds behaved normally, feeding the young bird mainly on deer and sheep carrion, supplemented with rabbits and birds. After fledging successfully in late July, the young Eagle was seen regularly with its parents in the territory throughout the remainder of the year, until last recorded in mid-December.

A second pair established a territory in the western edge of the county in 1976, some 32 km from the first pair, and started breeding.

From 1970 to 1981 the first pair reared a total of seven young and the second pair reared three young, though never more than one young was produced by either pair in any one year. Because of the high level of protection, the young birds were allowed to fledge without having been ringed or properly examined. Over the same period a number of Golden Eagles were found dead in various parts of the county, with no way of knowing where they came from or how old they were. It was therefore decided in 1982 to allow a single visit to each eyrie when the chicks were 5 or 6 weeks old, to ring and examine the young, and check the nest contents. On these first visits, it was discovered that the single chicks at both sites were slightly deformed. In the Haweswater nest, the upper mandible of the chick was twisted to the left, overlapping the lower mandible. The chick in the west of the county was similar – its upper mandible was even more skewed, but to the right. To date there has been no satisfactory explanation for these deformities. (Further discussion is given by the writer in Horne, 1994). Both young birds fledged successfully in July and were seen with their parents in their respective territories throughout the remainder of the year.

Unfortunately in the spring of 1983 the female of the second pair was found dead, having possibly been shot. The male also disappeared soon afterwards. Since then this territory has remained untenanted.

The Haweswater pair has continued to breed successfully since 1982 rearing a further 7 young, all of which were ringed and on examination found to be free of deformities. A total of nine chicks have been ringed in Cumbria, and one subsequently recovered dead. This was in 1986 when the young bird of the year was found in December, near Grassington, North Yorkshire, having been electrocuted on high voltage overhead cables.

The Haweswater Eagles begin their courtship in February with some spectacular aerial displays and bouts of nest building. The nest is usually fairly large, measuring approximately 1.2 to 1.5 metres in diameter and the same in depth. It is constructed from stick and heather stems, and lined with woodrush and grass. Laying takes place during the last week of March, which is relatively late compared with Golden Eagles in Scotland. The two eggs are laid at 4-day intervals and incubated by both birds, although the female takes the greater share. There are no definite records of both eggs ever hatching and the single chicks have taken between 11 and 12 weeks to fledge. Both parents feed the young, which stay with them in the territory until they become self-sufficient. In November or December the young leave the breeding territory.

Over the years 1993-1995 none of the eggs hatched, and examination showed that there was no embryonic development. With the resident birds now ageing (the male is estimated to be some 20 years old), there were suggestions that one or other partner had become sterile. This was disproved with successful hatching in 1996. Although breeding in Galloway is decreasing, recent years have seen the establishment of Golden Eagles in the Borders; the prospect of younger birds coming in to supersede or supplement the existing Lakeland pair is therefore still good.

Deliberate destruction of the 'preferred' nest in December 1989 (Walker, 1991), and again in early 1996, was doubtless intended to force the birds to build a new nest in one of the alternative sites more vulnerable to human predation. This well illustrates the continuing pressures on the birds and their protectors alike.

Buzzard *(Buteo buteo)*

The Buzzard is now a relatively common resident throughout the county, occupying low fell country and woods as well as river valleys, up to an altitude of 450 metres.

This is in contrast to the situation a hundred years ago. Macpherson records a few pairs of Buzzards breeding in *'our mountain solitudes'*, the species having been exterminated from most of its lowland sites. He also remarked that nesting sites tended not to be in trees, but in more inaccessible sites.

At the beginning of this century numbers were able to expand as a result of the reduction in the numbers of gamekeepers during and after the First World War; in addition a marked shift in range has also been noticed as birds follow the eruptions of Rabbits in lowland areas of the county. Blezard (1943) reported

Buzzards as re-established in the lower Eden valley in 1925, with an additional pair from 1926 onwards. Buzzards returned to the Mallerstang area in 1937, and in 1941 two pairs attempted breeding. The population in Cumbria in the 1920s is believed to have been between 300 and 400 pairs. This number remained stable up to the late 1950s when there was a dramatic decline in food supply due to the outbreak of Myxomatosis in the Rabbit population (Stokoe, 1962). Since this time the number of occupied sites has fluctuated with the abundance of Rabbits.

Birds are often seen during March in soaring display flight near their preferred nest site. A substantial nest of sticks or heather stalks, lined with moss or grass, is built during the first two weeks of April on a small crag or relatively high up in a tree against the main trunk. The placing of fresh green material on alternative nests may act as a territorial marker (Fryer, 1986). The 2 or 3 (occasionally 4) eggs are usually laid after the second week in April at two day intervals. Incubation starts with the first egg and takes between 28 and 30 days. Fledging takes between 6 and 7 weeks, during which time food is brought in to the nest by both parents. If at any time during the fledging period food is in short supply, the youngest of the brood may die or be killed and eaten by its siblings. In many cases where three chicks hatch, only one survives to fledging. Young stay with their parents for 5 or 6 weeks after fledging and are taught to fend for themselves. After this period they tend to disperse from the breeding area and seek territories.

Rabbit is the main food source, but other prey items recorded in the district include Mole, Common Rat, Field Vole, Wood Mouse, Brown Hare leverets, Carrion Crow, Jackdaw and Meadow Pipit (Stokoe, 1962). Buzzards have also been noted predating a large roost of Starlings and thrushes.

From late summer onwards, numbers of lowland birds are augmented by many (mostly juveniles) from the uplands, coming down to winter in the woods.

The National Buzzard Survey in 1983 (Taylor *et al.*, 1988) confirmed that the Cumbrian population had recovered to the extent that it was very nearly back to post-First World War numbers at approximately 380 pairs. Since then the population has expanded dramatically, following the increase in Rabbit numbers, to between 450 and 500 pairs. As the century closes, there is now a prospect of further devastation of Rabbit populations – this time through Rabbit Viral Haemorrhagic Disease. Any decline in this major prey species would not augur well for the Buzzard.

Sparrowhawk *(Accipiter nisus)*

The Sparrowhawk is a common resident, breeding in moderate numbers throughout the county. It inhabits scattered woodland and coniferous plantations, interspersed with cultivated land and open spaces to an altitude of 220 metres. Areas of pine and spruce plantation are favoured, although deciduous trees, particularly birch, are sometimes selected for nesting.

Sparrowhawks prey on small birds which they catch by stealthy approach and high speed attack. They can often be seen in suburban gardens or flying low along roadside hedges, slipping from side to side through gaps and accelerating to snatch small birds unawares. The flight speed in such attacks was measured by Brown (1974) as 48 km/hour. The surprise attack strategy of the Sparrowhawk appears high-risk, resulting in many fatalities to the hunter as well as the hunted. Sparrowhawk casualties due to crashing into windows are a frequent source of ringing recoveries. These are most often juvenile birds.

The species is remarkably sedentary, most birds generally remaining close to their nesting territories. Of 28 ringing recoveries detailed by Brown (*op. cit.*), the mean distance travelled (excluding the highest of 129km) was just 17.4km. Ringing by the writer over the period 1968-1993 involved 309 nestlings, of which 37 have so far been recovered, with average age at recovery of 20 months and mean distance travelled 19.2 km. A recovery from Ilkley, Yorkshire was by far the longest movement – 123 km. (Two birds ringed in Northumberland have been recovered in Cumbria, having moved 61 km and 106 km respectively).

The nest of small sticks and twigs is built against the main trunk of a tree, usually about two-thirds of the way up and always below the canopy level. Brown (*op. cit.*) stated the species nested impartially in coniferous or deciduous trees, with one nest only 12 feet up in a hazel bush. Eggs are laid at two-day intervals in early May. The clutch normally numbers four or five eggs, six is not unusual and seven is rare. Incubation is by the female alone, lasting thirty five days. The young take 24 to 30 days to fledge, during which time they are fed exclusively by the female on prey brought to the nest by the male. The female only begins to hunt for food once the young are 17 days old and able to feed themselves.

Despite persecution, the Sparrowhawk remained widespread in the county during the 19th century. In 1891 '*no fewer than eight nests of the Sparrowhawk were robbed of their eggs in the Carlisle district*' (M). However, as with most other raptors, there was a dramatic decline in Cumbrian numbers during the pesticide era of the 1950s and early 1960s, when toxic chemicals used as seed dressings also resulted in reduced brood sizes. Brown (*op. cit.*) reported that before 1955 32 Sparrowhawk broods had averaged 4.15 young per brood, but that after this date 16 broods had averaged just 2.18 young per brood, with four instances of just one young being reared (a situation not recorded before 1955).

With the voluntary restrictions on the use of DDT and dieldrin in 1962, 1964 and 1966, the population started to make a dramatic recovery. Newton and Haas (1984) gave figures for the percentage of 30 traditional nesting sites in Cumbria occupied over the period 1941–1978. These results showed an initial level of *ca* 85% occupation, which reached a nadir of 22% occupation in the pesticide years 1961-65, gradually recovering to 75% by 1978. By 1980 the population, largely freed from the effects of pesticides and persecution, had recovered to beyond the pre-1950 levels.

Since 1980 there has been a perceptible decline in the number of small birds, consequently reducing the Sparrowhawk population to the extent that it is once more below the pre-1950 levels. Some woods, particularly on the Solway Mosses, which consistently held pairs of birds during the pesticide era are no longer tenanted, and larger stands of plantation which held four or five pairs during the 1980s now hold only one. The future status of this bird in Cumbria is thus unclear.

Goshawk *(Accipiter gentilis)*

The fact that early naturalists, such as Heysham, made virtually no reference to the Goshawk suggests that by the end of the 18th century it was already long gone. But for Macpherson's diligence in finding a 13th century reference to an *'eyry of goshawks'* in *'Thomas' Wood in Bastonswayt'* [Bassenthwaite], there would be little to add to the reasonable supposition that the bird may have been present in the medieval woodlands.

There is a possibility that odd pairs may have bred in the early part of this century, but because of their secretive nature were overlooked. However, there is no real evidence of Goshawks breeding again until after they were re-introduced into the Border forests in the late 1960s and early 1970s. The re-introduction programme resulted in a very healthy population in these forests (mainly outwith Cumbria) and provided a basis for an expansion into other parts of the county.

A minimum of two territories had been established in central Lakeland by 1975 and breeding was suspected, although never properly confirmed, at these sites. The first confirmed breeding record of Goshawks outside the Border forests was in 1984, when a pair from one of these two territories reared 2 young. At present the number of breeding pairs in the county is between 6 and 8; three are breeding in the south of the county, another two in the north, and possibly a further two or three in the western valleys which contain suitable woodland. There is still sufficient suitable breeding habitat for Goshawks to increase their population further – the main inhibiting factor being the continued high level of persecution which this species attracts.

The breeding season starts in late March and early April when birds may be seen performing their display flight over the wood in which they intend to nest. The nest and its situation are very similar to that of the Sparrowhawk but, with a greater bulk of nest, building tends to occur in larger trees. Egg-laying begins in late April or early May with the eggs being laid at two day intervals. Incubation is entirely by the female alone and probably takes 36 to 38 days. Monitoring of the Border forest population in 1987-95 showed that clutch sizes averaged 3.42, with a mean brood size of 2.26 (Petty & Anderson, 1995).

Food is brought into the nest by the male and the chicks are fed by the female during the first three weeks of growth. Beyond this stage the female also hunts, and leaves food for the chicks to deal with. The young fledge after 42 to 43 days.

Prey items recorded from the Border forest population in 1975-82 were *ca* 95% birds; the main items (in decreasing frequency) were Feral Pigeons, Wood Pigeons, Red Grouse and corvids (Petty & Anderson, *op. cit.*).

Hen Harrier *(Circus cyaneus)*

Macpherson's comment in 1892 accurately reflected the fate of the Harrier: *'though gone from our midst as a breeding species, and totally unknown to most residents, an odd bird still appears occasionally in Lakeland on migration'*.

This was the more telling because almost exactly a hundred years earlier Dr John Heysham had been observing several breeding pairs on Newtown Common – now a suburb of Carlisle (Heysham, 1794). There is clear evidence that the great lowland Mosses, of the Solway especially, were then a stronghold of this bird. By the beginning of the present century it had been banished from the British mainland – breeding only in the Northern Isles.

Of breeding attempts in the earlier years of this century, there is only scant evidence. Blezard (1943) and Graham (1993) both refer to a breeding attempt in north Cumberland in the period 1925 to 1928. A Westmorland attempt in 1936 is also mentioned.

The relatively remote moorlands of the northern Pennines and Borders are probably now the only part of the county which can provide sufficient solitude for this species to breed. From the 1960s onwards, afforestation programmes have changed the nature of vast tracts of this region north of the Tyne Gap. Some of the newly planted ground proved attractive to the Harriers – and may have provided useful alternatives to the more hostile game preserves. Since 1978 one or two pairs have regularly reared young, but at the present time there are at least 5 or 6 pairs attempting to breed each year, with mixed success.

Birds usually arrive back on their breeding territory during April and eggs are laid in the first half of May in a rudimentary nest on the ground, built from heather fronds or dead rushes. The 4 or 5 eggs, occasionally 6, are incubated by the female alone for approximately 30 days. During this time the male provides all of the food and passes it to her in a spectacular aerial pass. The young take between 5 and 6 weeks to fledge. It is at this time that they are at their most vulnerable; foxes will take and kill young in the nest, but human predators are also a significant problem, despite full legal protection.

An incident at the Geltsdale Reserve near Carlisle in 1989 provided a sombre reminder of the latter: two nests, both with 6 young, were destroyed. One nest was found empty, but clearly disturbed; at the other all the young were dead, with heads and wings pulled off. A subsequent court action did not yield a conviction.

With the present dramatic increase in the population in southern Scotland, and the large number of young being reared there, the outlook for Hen Harriers in Cumbria is potentially good but, as the above incident proves, the future is still heavily dependent on those who own or manage its breeding grounds.

Peregrine Falcon *(Falco peregrinus)*

'No species could increase in the face of such destructive measures; but up to 1878 a few pairs continued to breed in the east of Cumberland and East Westmorland, but especially in the heart of the Lake District.'

Macpherson's account, from which the above is quoted, paints a dismal picture of a species struggling against the combined forces of game preservers, egg collectors, falconers and specimen hunters. The Peregrine, even in Lakeland, was a scarce bird in the 19th century – though no accurate figures can be given.

Ernest Blezard (1943) summarised the situation in the counties of Cumberland and Westmorland half a century later as consisting of *'nearly 20 pairs'*. Further work by Derek Ratcliffe in the 1950s and 1960s revealed that this figure was an under-estimate and that the population was more likely to have been between 35 and 37 pairs (Ratcliffe, 1993). The creation of Cumbria in 1974 added parts of North Yorkshire to the area previously considered, bringing a further 3-4 breeding pairs into the reckoning. This 'new' area, with a grand total of 41 pairs in 1940, is taken as the baseline against which trends discussed below are measured. Undoubtedly the figure of 41 pairs was artificially low due to the still relatively high level of persecution of Peregrines in the first half of the century. In addition to the continuing pressures of the previous era, homing-pigeon fanciers were also instrumental in keeping the numbers of birds down.

Historically, the main concentration of territories throughout the county was in central Lakeland on land over 350 metres and up to 600 metres altitude. Traditional nesting crags of 'first class' status were usually very precipitous faces on average 60 metres high, mostly fairly remote from human habitation and mostly facing between east and north – although this may in part be a quirk of the Lakeland landscape. These were also the territories which held the few remnant birds during the toxic chemical poisoning episode of the 1950s and 1960s.

The pesticide effects were first noticed nationally in the 1950s, and by the early 1960s the Cumbrian population had 'crashed' to 8 pairs (an 81% decrease). Of these only 3 or 4 pairs were rearing young in any one year. At this stage the county's Peregrine population was on the verge of extinction.

Following the restrictions on the use of pesticides in 1962, 1964 and 1966, signs of recovery were noticed by 1967. Increases from such a low level are likely to have involved movements into the county from other areas, particularly the Highlands of Scotland, which were the least affected by such influences. The national Peregrine survey of 1971 (Ratcliffe, 1972) established that in Cumbria the number of pairs holding territory was 21 – just over half the pre-pesticide baseline figure.

During the initial stages of the recovery period the first territories to be re-occupied were the traditional 'first class' crags. The new birds instinctively returned to exactly the same ledges and even to the same nesting scrapes which had been used by past generations of Peregrines. The situation continued to improve throughout the whole of the county and by 1976 the number of occupied territories (42) actually exceeded the baseline figure.

Up to the next national survey in 1981, it was obvious that numbers were still increasing and that new 'second class' territories were being established. These tended to be centred on smaller, less remote, crags, as well as in both working and abandoned quarries. Recruitment to the Cumbrian population during the 1970s was approximately 3.5 new pairs annually. The 1981 survey figures showed that whilst the national population had made a 90% recovery, the Cumbrian population, now with 68 occupied territories, had increased by 165%. During the 1980s new pairs continued to appear, now often using 'third class' eyries. These included sites which had no previous history of occupation by Peregrines, such as small rock outcrops in lowland river valleys, and industrial slag banks very close to human habitation. As in the previous decade, the population in Cumbria continued to increase by about 3.5 pairs annually. By 1991, it had reached a total of 94 occupied territories, a 229% increase. This remarkable achievement gave the county the highest density of Peregrines in the world. Since 1991, the population has further increased to a total of 98 occupied territories, which is generally regarded as saturation level. Appendix 1 illustrates this recovery graphically.

From ringing returns it is obvious that Cumbrian birds have already been instrumental in the re-colonisation of other parts of the country. Ringing undertaken by the writer from the late 1960s onwards has now exceeded 1000 fledglings, 110 of which have subsequently been recovered. The three oldest individuals were all in their 13th year and all found within the county. Appendix 3 maps the dispersion of these recoveries.

The Peregrine's breeding season usually starts in early April, although the odd pair can have a full clutch of eggs before the end of March. The average laying date for the county is 12th April, but in years with extremely cold wet spring weather, this can be delayed by one or two weeks. The 3 or 4 (rarely 5) eggs are laid at two day intervals in a rudimentary scrape on a grassy cliff ledge. On occasions an old Raven nest, which gives a greater amount of protection from the weather, will be used. The 31-day incubation starts with the third or fourth egg and is mainly by the female. Fledging takes place at 5 to 6 weeks and the young stay with their parents for a further two months, at which stage they can fend for themselves and wander – or peregrinate – as they please. Appendix 2 gives a summary of breeding performance over the past 30 years.

The Peregrine has in many ways provided a classic conservation success story – though not without great effort on the part of many people. A local dimension to this relates to Falcon Crag, Keswick, where egg and chick thieving was renewed when the site was re-occupied in 1970. To combat this, members of Keswick Natural History Society and the Cumbria Wildlife Trust provided continuous daytime surveillance, and Keswick Mountain Rescue team slept under the crag at night. The success of this operation has been a tribute to popular concern for wildlife. 4 young were reared in the first season (1981), and 37 have now fledged since these operations began.

It is of course impossible to safeguard all sites to this degree. At the time of writing, a poisoning incident at a west Cumbria nature reserve has been attributed to pigeon racing 'activists'. Whether or not this is correct, it is sadly the case that the Peregrine, like many other raptors, may always be seen as an unacceptable competitor by a few who are prepared to flout the law.

Merlin *(Falco columbarius)*

Merlins are partial residents in Cumbria, nesting on heather-clad moorland and hillsides, in trees in upland areas, and occasionally on coastal sand dunes among marram grass (Blezard, 1943). They formerly nested on some of the Solway Mosses. In recent years there has been a tendency to move into the edges of forest plantations, which have become suitable breeding habitat.

Merlins usually leave their moorland haunts after their prey species depart during mid-August to October, and can often be seen along the coast in winter, hunting small waders and passerines. They arrive back on breeding territory from late February to mid-April or early May. The nest is a rudimentary scrape on the ground amongst rank heather, sometimes with small pieces of heather as lining. They will also use old crow nests in isolated trees on the moor or in conifers along the edge of forestry plantations and have been reported using an old Buzzard nest (Blezard, *op. cit.*).

Nesting on the ground, Merlins are prone to predation by Foxes, Stoats and Weasels – and in some situations Buzzards will take the young as prey. In addition, with the increase in numbers of Peregrines, some birds have been forced off territory, resulting in a decline in the number of sites occupied. Eggs are laid in early May, but occasionally into June. A clutch of four is normal, but sometimes five and rarely six have been found. The young may fledge as early as the beginning of July, but the second or third week of that month is more usual (Brown, 1974). The commonest brood size reported by Brown was 3, with an average of 3.16 young per brood from a sample of 19. Merlins may disperse some distance: Blezard (*op. cit.*) records ringing recoveries of Cumbrian fledged birds in Yorkshire, Lancashire, Shropshire and Dax, (Landes) France.

By the end of the 19th century the Merlin had succumbed to the relentless persecution of the time and Macpherson feared it would soon be extinct as a breeding species. He lamented the pointless destruction of this bird, which certainly was not guilty of the damage to grouse stocks of which it stood accused. Macpherson is supported by Brown (*op. cit.*) who stated that he had never seen any prey remains of game birds (adults or young) at the nests or plucking sites of Merlins over a fifty-year period. Dunlop (1923) made only one reference to this species – as having nested at Langdale (where the prey items were Swallows).

Blezard (*op. cit.*) reported Merlins to be sparingly distributed over the fells (most frequently in the Pennines) and some low-lying mosses. Brown (*op. cit.*) stated that they had ceased breeding on the mosses from 1930. Stokoe (1962) reported a decline in breeding pairs over the previous 10 years.

In 1993/94 a national Merlin survey appeared to demonstrate a slight increase in the Cumbrian breeding population over the 10-year period since the previous survey, in line with national trends. This was certainly true of the Pennines, where all sites used in 1983/84 were still occupied, with several new pairs in addition (Shackleton, 1996). The situation in the Pennines contrasts with other areas in the Lakeland fells where traditional breeding territories are being vacated as heather cover decreases through over-grazing. In 1995 just 32 out of 116 traditional Cumbrian breeding sites were known to be occupied. It appears that the present population lies between 30 and 40 pairs and that breeding performance has been poor in recent years. The declines in some parts of Cumbria contrast with Northumberland and Durham, where numbers are increasing on the moors and along forest edges. While the Merlin appears to be holding its own in the Pennines, its fate in the Lake District will be closely linked with the management of sheep-grazing and a reversal of the loss of heather cover.

Kestrel *(Falco tinnunculus)*

As the commonest bird of prey in Cumbria, the Kestrel can often be seen in towns and, more frequently, along motorway verges and cuttings, or perched on roadside trees or posts. It has long been regarded as widespread and a common resident, breeding in good numbers. It is extremely difficult to put a figure on the population and, not unexpectedly, the BTO surveys of 1968-72 and 1988-91 revealed its presence in every 10 km grid square in the county.

This may well also have been the case a century ago, since Macpherson comments little on its status. That the species had not entirely escaped the treatment meted out to other predatory birds is suggested by his remark that *'Public opinion has begun to recognise that the injury which the Kestrel inflicts on game preserves is exceedingly small'*.

The abundance of the bird is doubtless made possible by its tolerance of man, and adaptability to a wide range of habitats. In Cumbria it breeds from sea level up to an altitude of *ca* 600 m. The sight of birds hunting over open moorland is frequent: newly forested upland can be a particular attraction when voles are abundant. This was noted by Ritson Graham many years ago (Graham, 1993) and has been equally true over the new Border forests of the 1960s and 1970s. Kestrels have been seen sharing hunting territory with Hen Harriers, Short-eared Owls and Barn Owls simultaneously in such terrain. They have been regular breeders on the sea cliffs at St Bees Head, and have occupied man-made 'cliffs' such as Carlisle Cathedral and other urban buildings. Although largely resident, several observers have noted that Kestrels tend to move away from the uplands in winter. Referring to the Bewcastle fells, Graham (1993) remarked on a *'definite arrival of nesting pairs'* each spring.

Ringing carried out by Brown (1974) produced mainly local recoveries, but included two from northern France. Another bird, which had travelled south to Derby in its first year, was back close to its site of origin in Cumbria 3 years later.

Kestrels lay in a scrape in a hole on a small cliff or building and in lowland areas a flattened old crow's nest may be used. The 4 or 5 eggs are laid from the middle of April onwards. Incubation of the eggs is by the female alone and hatch in about 28 days. The female broods the young chicks and feeds them for the first 14 days on food brought in by the male, after which both birds will hunt for food and feed them separately. The young fledge after 28 days and stay with the adults until they are able to hunt and fend for themselves. Performance is governed by the availability of the main prey species, the Field Vole, with larger broods in good 'vole years'. Insects also feature abundantly as prey items, and Kestrels frequently take such prey on the wing. An analysis of pellets from a Pennine valley by Blezard (1954a) identified many ground-dwelling beetles amongst the remains, which included Carabids and Geotrupids.

In some areas of the county there are recent reports of a declining number of Kestrels over the past 5 or 6 years, perhaps due to a series of cold wet springs and fewer voles.

Raven *(Corvus corax)*

Raven place-names are frequent in the area, especially in the rugged hills of the Lake District, where 'Raven Crag's are almost commonplace – at least 33 being known. This amply testifies to the long history of the bird as a breeding resident in the county, which remains the main stronghold of the Raven in northern England.

Heavily persecuted in earlier times, Ravens were (and still are) accused as killers of lambs and sickly sheep. In some Parishes, bounties were once in force. The poet Wordsworth, writing in 1805, recollected *'frequently seeing, when a boy, bunches of unfledged Ravens suspended from the churchyard at H____ , for which a reward of so much a head was given to the adventurous destroyer'*. Macpherson conjectured that H refers to Hawkshead.

The frequency of 'fallen stock' on the fells at most seasons means that sheep-killing is scarcely a necessary strategy for Ravens. Whilst sheep (and more locally deer) carrion is their staple diet throughout most of the year, they also feed on sick or injured birds and mammals as well as the large black slug *Arion ater*. Blezard (1954a) reported remains of a Field Vole and several dor beetles (*Geotrupes* sp.) in a small sample of pellets, in addition to sheep's wool and bones.

With declining persecution, the Cumbrian Raven breeding population has probably been relatively stable for much of the century. Blezard (1954b) revised an earlier estimate of *'nearly 40 pairs'* to 60 pairs. More recent estimates by the writer suggest that 80-90 pairs is a more accurate total nowadays. While the earlier revision reflected improved survey coverage, the later one arguably points to a real population increase. The majority of these birds are in central Lakeland on land above 300 metres.

There were formerly 5 or 6 pairs in the northern Pennines, but these had disappeared by 1988 due to higher levels of persecution in that region. Blezard (1928) stated that of nine known Pennine nesting sites *'rarely, if ever, are more than two . . . occupied in one season . . .'* : this also is now considered to be an underestimate due to difficulties of monitoring.

Sea-cliff nesting at St Bees Head was first reported this century in 1947 (Blezard, 1954b) and has continued to the present day. There has been one other sea cliff site in very recent years. Quarry sites have also been used, and more recent instances may possibly reflect new pressures from the burgeoning Peregrine Falcon population.

In spring, Ravens can often be seen soaring and gliding around their home crag, performing an impressive synchronised display in which the two partners fly very close together, matching each other's movement and direction perfectly, turning onto their backs for short distances and at regular intervals. Their habitat is shared with Peregrines, and where both nest fairly close together, or compete for the same site, they are bitter enemies. Aerial fights can be quite spectacular, with the falcon stooping at the Raven, which at the last minute turns onto its back, talons at the ready, to defend itself. Undoubtedly, the superior power of the Peregrine ensures that it normally has the better of such encounters.

The Raven usually builds a large nest under an overhanging rock on the most precipitous section of the crag. Tree-nesting has occurred relatively infrequently, possibly reflecting a greater vulnerability to persecution. The nest is constructed from heather stalks with a warm lining of moss and sheep's wool formed into a deep cup. Egg-laying usually takes place during the first two weeks of March, although some birds can have a full clutch of 5 or 6 (rarely 7) eggs before the end of February. Erythristic (pink) eggs (instead of the normal blue/green) have been found on a number of occasions, though not over the past twenty years. Raven eggs are relatively small for the size of the bird, which possibly explains the very short incubation period of 20 to 21 days.

Clutch sizes in northern England during 1935-1995 averaged 4.91 (153 clutches); fledged broods in the period 1935-1975 averaged 2.80 young (91 broods), but showed a significant increase to 3.11 young (256 broods) over the years 1976-1995 (Ratcliffe, in prep.).

Whilst the Raven can usually cope with severe weather during the breeding season, there have been some years when the weather at high levels on the hills has been extreme, resulting in up to 40% of first clutches of eggs failing or small young dying in the nest. If the first clutch of eggs fails (or it is taken by collectors), the birds will usually lay a repeat clutch after a period of approximately three weeks. Fledging takes place after a period of 6 weeks and the young birds stay with their parents and other Raven families well past the end of the breeding season.

Winter flocks, foraging for food and coming together into communal roosts, are well known. Blezard (1954b) gives useful detail of this behaviour and quotes early records of large Pennine roosts, some amounting to over 50 birds. It is likely that the bulk of these will have originated from the Lake District: regular (even daily) movements of Ravens between this area and the Pennines have been noted throughout the century. Such movements tie in with Brown's comment on recoveries of ringed Ravens indicating '*a local movement . . . into the Yorkshire Dales and moors*' (Brown, 1974). It is also unfortunately the case that the reception given to Ravens on some of the Pennine moors may have produced an un-natural bias towards such eastern 'recoveries'. Although Pennine roosts no longer reach the large numbers of earlier years, winter roosts and flocks in the Lake District can still contain 50 or more birds.

Derek Ratcliffe's study of this species (Ratcliffe, in prep.) includes much information gleaned from work on Ravens in Cumbria over many years. It also reveals the character and resourcefulness of this bird which, despite the pressures mentioned above, is likely to remain 'the spirit of the fells' for many more centuries to come.

Concluding remarks

The century has seen important changes of public attitudes to birds of prey and to wildlife in general. The outlook for the predatory birds is in many ways now more hopeful than at any previous time during the past one hundred years. Despite this, some of the 'traditional' threats still remain, and yet others have developed. As the pesticides experience proved, problems not previously anticipated may develop quickly, demanding unceasing vigilance from all concerned.

The relatively precarious situation of the Golden Eagle as a breeding species has already been referred to. One of the more optimistic prospects is that species extinct as breeders for well over a century may soon be back in the county. The recent expansion of the Osprey in Scotland augurs well for its return – certainly it is now seen in increasing numbers on passage; the Red Kite is currently extending its British range, in part through re-introduction programmes elsewhere in the country.

It remains to be seen whether these and other species can successfully adapt to the ever-shifting pressures which the next decades will inevitably bring.

Acknowledgements

My indebtedness to Ernest Blezard, Robby Brown, Teasdale Stephenson and Ray Laidler for their friendship, guidance and encouragement in my fieldwork during the 1960s and 1970s is immeasurable. I shall always be grateful and consider it a privilege to have known them and been able to learn so much from their vast store of knowledge of natural history. I am most grateful to Derek Ratcliffe for all his help, encouragement and guidance. His companionship in the field is much appreciated, as is that of two stalwart friends John Davidson and Bob Buchanan, who were always there to encourage and assist when needed.

I wish to record my gratitude to Colin Armitstead, Geoff Fryer, Derek Hayward, Terry Pickford and Paul Stott for generously and freely sharing the results of their work for other areas of the county.

Much help has also been received from RSPB staff at Haweswater and Geltsdale over many years, especially John Day, Dave Shackleton, Malcolm Stott, Stephan Ross and Dave Walker.

Many friends and associates throughout the north of England have been generous with their help, especially the following: David Anderson, Ian Armstrong, Dorothy Blezard, John Callion, Mike Carrier, Peter Davies, Martin Davison, Ian Findlay, Robin Griffiths, John Hamer, Neil Henderson, Steve Hill, Ken Hindmarsh, Dorothy and Stewart Illis, David Jardine, Colin Jewitt, Tony and John Laidler, Brian Little, Paul Marsden, Paul Martin, Stephen Martin, Alan McKenzie, John Miles, Mick Mills, Steve Petty, Jean Scott, and Terry Wells.

Finally, I am most grateful to the Editors, David Clarke and Stephen Hewitt, for additional historical material and for their forbearance and assistance in the compilation of this paper, and also to Dr John Todd who kindly provided useful references to medieval documents.

References

Blezard, E., (1928), On the Raven. *Transactions of the Carlisle Natural History Society* **IV**: 16-22.

Blezard, E., (Ed.), (1943), *The Birds of Lakeland*. Transactions of the Carlisle Natural History Society **VI**.

Blezard, E., (1946), *Lakeland Natural History*. Transactions of the Carlisle Natural History Society **VII**.

Blezard, E., (1954a), Food of birds. Pp.75-101 in *Lakeland Ornithology*. Transactions of the Carlisle Natural History Society **VIII**.

Blezard, E., (1954b), The birds of Lakeland: a second supplement. Pp. 102-13 in *Lakeland Ornithology*. Transactions of the Carlisle Natural History Society **VIII**.

Blezard, E., (1958), *Lakeland Birds*. Transactions of the Carlisle Natural History Society **IX**.

Brown, R.H., (1974), *Lakeland Birdlife 1920-1970*. Carlisle: privately published.

Cumberland Nature Reserve Association, (1915), 1st Annual Report 1914. *The Naturalist* **June/July 1915**: 189-243.

Dunlop, E.B., (1923), Lakeland Ornithology 1892-1913. *Transactions of the Carlisle Natural History Society* **III**: 1-39.

Fryer, G., (1986), Notes on the breeding biology of the Buzzard. *British Birds* **79**: 18-28.

Gibbons, D.W., Reid, J.B. & Chapman, R.A., (1993), *The New Atlas of Breeding Birds of Britain and Ireland 1988-1991*. London: T. & A.D. Poyser.

Graham, R., (Eds: Matthews, S. & Clarke, D.J.), (1993), *A Border Naturalist: the birds and wildlife of the Bewcastle Fells and Gilsland Moors 1930-1966*. Carlisle: Bookcase.

Heysham, J., (1794), A Catalogue of Cumberland Animals. Pp. 1-38 in Hutchinson, W., *The History of the County of Cumberland* Vol I. Carlisle: S. Jollie.

Horne, G., (1994), Mandible deformities in raptors. *Birds in Cumbria* **Spring 1994**: 74-76. Cumbria Naturalists Union.

Macpherson, H.A., (1892), *A Vertebrate Fauna of Lakeland*. Edinburgh: David Douglas.

Newton, I. & Haas, M.B., (1984), The return of the Sparrowhawk. *British Birds* **77**: 47-70.

Petty, S.J. & Anderson, D.I.K, (1995), *Goshawks in the Border Forests in 1994 and 1995*. Confidential report of the Forestry Commission Research Division.

Ratcliffe, D.A., (1970), Changes attributable to pesticides in egg breakage frequency and eggshell thickness in some British birds. *J. Appl. Ecol.* **7**: 67-115.

Ratcliffe, D.A., (1972), The Peregrine population of Great Britain in 1971. *Bird Study* **19**: 117-156.

Ratcliffe, D.A., (1993), *The Peregrine Falcon*. 2nd edn. Calton: T. & A.D. Poyser.

Ratcliffe, D.A., (in prep.), *The Raven*. London: T. & A.D. Poyser.

Shackleton, D. (1996), Merlins in Cumbria – status and trends. *Birds and Wildlife in Cumbria* **Spring 1996**: 74-77. Cumbria Naturalists' Union.

Sharrock, J.T.R., (1976), *The Atlas of Breeding Birds in Britain and Ireland*. Tring: British Trust for Ornithology.

Stokoe, R., (1962), *The Birds of the Lake Counties*. Transactions of the Carlisle Natural History Society **X**.

Taylor, K., Hudson, R. & Horne, G., (1988), Buzzard breeding distribution and abundance in Britain and northern Ireland in 1983. *Bird Study* **35**: 109-18.

Walker, D.G., (1991), *The Lakeland Eagles*. Penrith: privately published.

Appendix 1

Peregrine Falcon: the 'super-recovery' in Cumbria, following the decline induced by pesticides in the 1950s and early 1960s. (Refer to text and Appendix 2.)

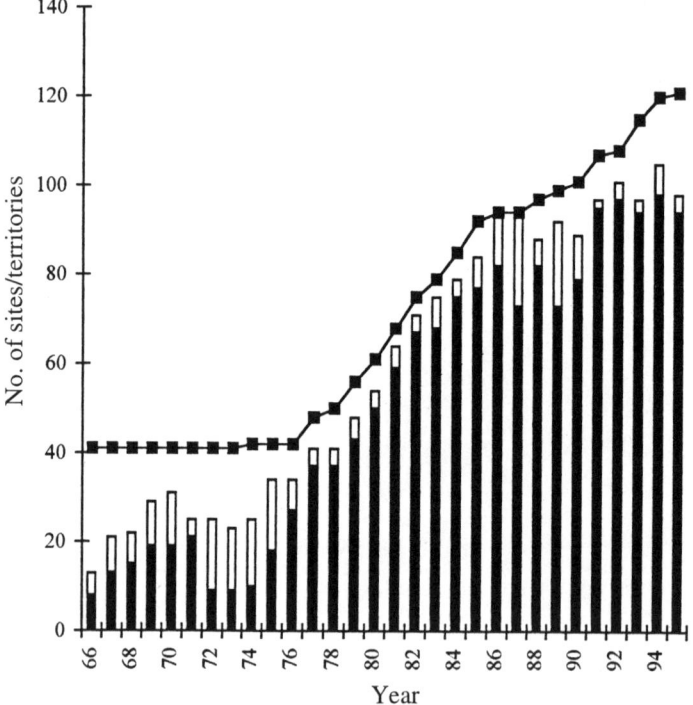

Solid column areas: occupied sites – verified by monitoring visits
Open column areas: occupied sites – inferred by calculation
Line graph: number of known territories

Appendix 2
Breeding performance of Peregrine Falcons in Cumbria 1966 – 1995
Data gathered from fieldwork by the writer and co-workers.

	1966	1967	1968	1969	1970	1971	1972	1973	1974	1975
Known territories*	41	41	41	41	41	41	41	41	42	42
Territories examined	25	25	28	27	25	35	15	16	17	22
Occupied territories	8	13	15	19	19	21	9	9	10	18
Pairs rearing young	2	7	7	11	7	10	2	6	4	10
Young reared	4	18	13	21	15	28	3	16	11	22
Mean brood size	2.00	2.57	1.86	1.91	2.14	2.80	1.50	2.67	2.75	2.20
Mean young per pair	0.50	1.38	0.87	1.11	0.79	1.33	0.33	1.78	1.10	1.22

	1976	1977	1978	1979	1980	1981	1982	1983	1984	1985
Known territories	42	48	50	56	61	68	75	79	85	92
Territories examined	33	43	45	50	57	63	71	72	81	84
Occupied territories	27	37	37	43	50	59	67	68	75	77
Pairs rearing young	9	18	21	19	27	23	44	37	44	39
Young reared	22	42	51	45	72	50	121	81	113	89
Mean brood size	2.44	2.33	2.43	2.37	2.67	2.17	2.75	2.19	2.57	2.28
Mean young per pair	0.82	1.14	1.38	1.05	1.44	0.85	1.81	1.19	1.51	1.16

	1986	1987	1988	1989	1990	1991	1992	1993	1994	1995
Known territories	94	94	97	99	101	107	108	115	120	121
Territories examined	83	73	90	79	90	105	104	111	112	116
Occupied territories	82	73	82	73	79	95	97	94	98	94
Pairs rearing young	39	38	47	43	44	64	53	41	62	57
Young reared	88	95	110	106	116	153	125	85	148	150
Mean brood size	2.26	2.50	2.34	2.47	2.64	2.39	2.36	2.07	2.39	2.63
Mean young per pair	1.07	1.30	1.34	1.45	1.47	1.61	1.29	0.90	1.51	1.60

Mean brood size is number of young reared per number of pairs rearing young.
Mean young per pair is number of young reared per number of occupied territories.

* *ie* the historically accumulated total of known breeding sites.

Appendix 3 (a)

Recoveries of adult Peregrine Falcons ringed by the author in Cumbria as nestlings 1975 – 1995

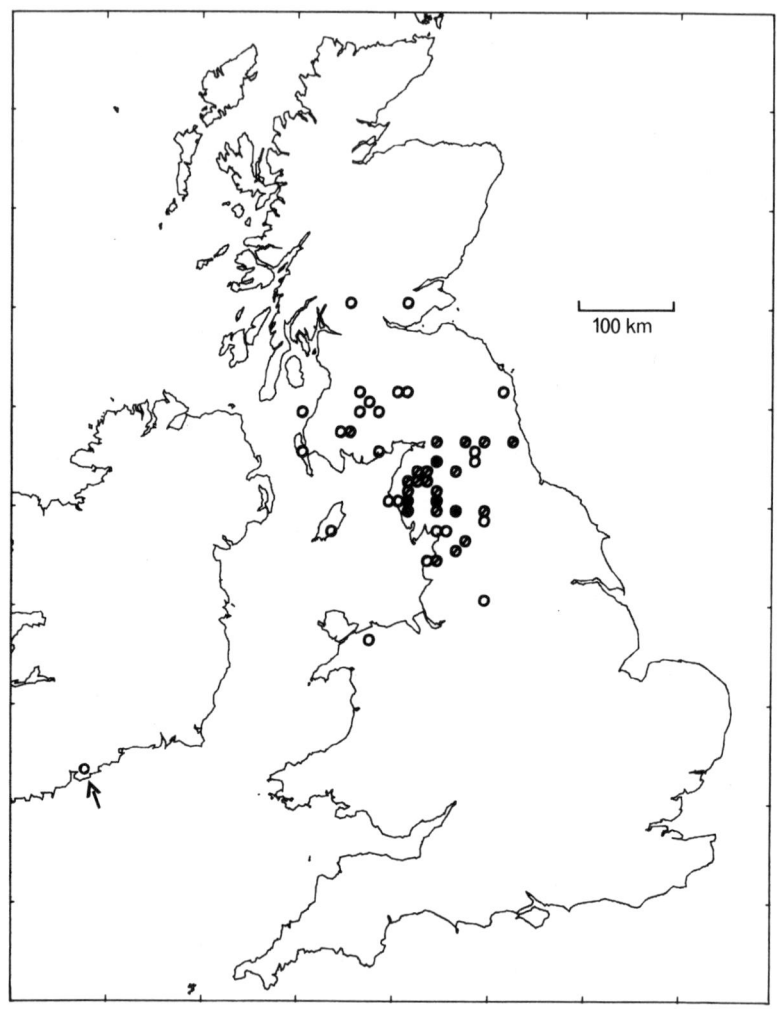

○ adult female
⊘ adult male
● adults – both sexes

n = 30 (female); 26 (male)

Birds of prey in Cumbria

Appendix 3 (b)

Recoveries of first-year* Peregrine Falcons ringed by the author in Cumbria as nestlings 1975 – 1995

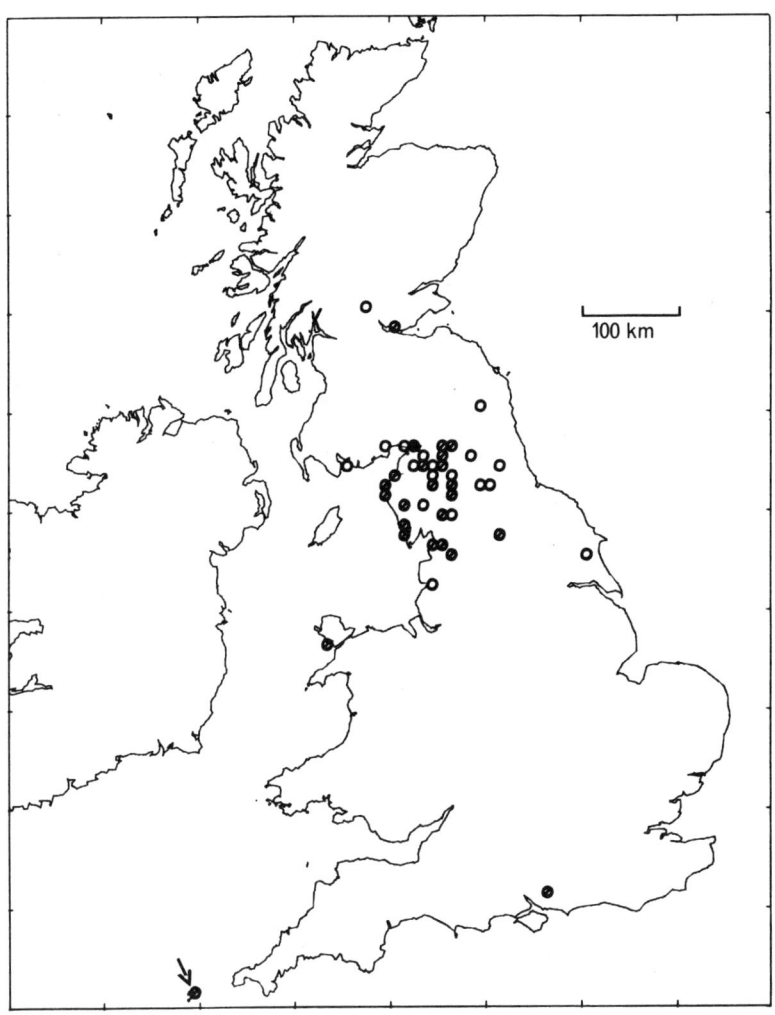

○ first-year female
⊘ first-year male
● first-year – both sexes
n = 19 (female); 28 (male)

* = *bird recovered before May of year following hatching*

MAMMALS IN CUMBRIA – a centenary review

John Webster

Obvious to all naturalists are the habitat changes that have occurred in our county this century: loss of deciduous woodland; conversion of moorland, moss and fell to conifers; the intensification of farm-land management; the increase in road construction and tourism. All have had their influences on wildlife, but there are two reasons, not unconnected, why we can still present a report on the mammals of Cumbria that is more encouraging today than could have been offered in 1893. Firstly, there has been a remarkable growth in our knowledge of British mammals in the last 40 years, and secondly, in the last 20 years, an increased and enlightened new level of protective legislation and conservation. The simple fact is that there seems no evidence of loss in the county's total number of mammal species, and although some populations are reduced, others are greatly increased.

My method in this paper will be to follow the order of the species used by H.A. Macpherson in his *Vertebrate Fauna of Lakeland* (1892), and review very briefly those where changes of status have been slight. I shall give more space to others where changes have been greater or have some additional element of particular interest.

As a group, bats reveal well the increases in knowledge and protection to which I have referred. The public's attitude towards bats has changed fundamentally owing to publicity on their behalf, publicity which followed a national decline in bat numbers this century – thought to be the result of habitat loss, pesticide use and timber treatments. County Bat Groups provide advice and assistance to the public, erect boxes to provide roost sites and monitor populations. Cumbria's first Bat Group was formed in 1983 and today the county is covered by the Westmorland and Furness Bat Group and the Cumberland Bat Group. All species and their roosts have been totally protected in Britain since the Wildlife and Countryside Act of 1981.

Of the species recorded in the county, only the Barbastelle *(Barbastella barbastellus)* – reported but once from the nineteenth century – is not known today. However, our number of species remains the same – at seven, as it is now known that Brandt's Bat *(Myotis brandtii)* is a species separate from the Whiskered Bat *(Myotis mystacinus)* with which it was once confused.

The Long-eared Bat *(Plecotus auritus)* Macpherson regarded as *'less evenly distributed than the Pipistrelle, but in some places... almost as numerous'*. Now called the Brown Long-eared Bat, it retains this status as the second commonest species. With a preference for roosting in large, pitched roofs, it is known across the county.

The Noctule *(Nyctalus noctula)* was not thought by Macpherson to be generally distributed, and he records only one site – at Bowness-on-Solway – where he thought they were present, in 1888. Numbers apparently have not been high this century, but the species has been seen in many parts of the county, and several more roosts have been located. In contrast, Macpherson thought the Pipistrelle *(Pipistrellus pipistrellus)* 'fairly common' and 'tolerably abundant'. This is one species that has perhaps benefitted from the twentieth century loss of woodland, and additionally from modern housing which it favours for roosts. This is our commonest species, with a wide distribution and one or two exceptional roosts containing around a thousand individuals.

It seems likely that Daubenton's Bat *(Myotis daubentoni)* has hardly changed in status. Always somewhat local on account of its feeding association with water, it is probably widespread although under-recorded. Recent survey work on the bridges of the county has revealed several previously unknown roosts of this species (Norman, 1995). Natterer's Bat *(Myotis nattereri)*, known to Macpherson as the Reddish-Grey Bat, was recorded by him from two sites only in Cumberland. It has now been recorded from a few localities across the whole of Cumbria. The Whiskered Bat, Macpherson knew from a few specimens mainly from along the Eden Valley, and this species is now thought of as widespread, but thinly distributed, while the closely related Brandt's Bat seems to be distinctly scarcer. It has, though, been found in the north of the county, and one large nursery roost of eighty bats at Spark Bridge in the south was a new find in 1992. There are few other records.

So knowledge of our bats continues to accumulate through the activities of the countys Bat Groups and their monitoring of boxes. It is hoped that Leisler's Bat *(Nyctalus leisleri)*, may soon be added to the county list, as records have recently come from Burton-in-Lonsdale, just south of the county boundary, and from Galloway to the north.

From the bats we turn to the five terrestrial insectivores recorded by Macpherson. They are all still present. The Hedgehog *(Erinaceus europaeus)* has a distribution related to that of scrub and deciduous woodland cover owing to the importance of leaves for the winter hibernaculum, so its presence in the county is sometimes localised. It is highly susceptible to modern levels of motorised road traffic, but deliberate persecution has largely stopped. The Hedgehog has protection under Schedule VI of the Wildlife and Countryside Act as do the three shrews, Common Shrew *(Sorex araneus)*, Pygmy Shrew *(Sorex minutus)* and Water Shrew *(Neomys fodiens)*. The last of these is not so common as the others as its habitat is more restricted – mainly to the banks of fast-flowing, unpolluted rivers and streams. The fifth insectivore, the Mole *(Talpa europaea)*, is still common, being highly adaptable and present in most habitats where the soil is deep enough to allow tunnel construction and there are sufficient earthworms and insect larvae as prey. This animal is still mercilessly persecuted as an agricultural pest, and often offensively displayed as a mark of the trapper's prowess.

The sad situation of the carnivores by 1890 is strongly brought out by Macpherson. He writes, *'It is clear from parish records that in the seventeenth century the dalesmen developed a murderous propensity for slaughtering all the wild animals whose presence supplied a charm to their . . . mountains'*. Hunting, trapping and shooting had by the end of the last century brought many populations of the carnivores to generally low levels.

This indeed could not be said about the Fox *(Vulpes vulpes)*, which even then provided good sport for the huntsmen of Cumbria, despite a list of bounty payments extending back over 250 years. Today the Fox remains without legal protection and, although hunted, snared, shot, gassed, poisoned and 'lamped', is common throughout the county. One Kendal lamper alone anticipates he will kill 200 each winter.

The Pine Marten *(Martes martes)* is, and was, in the opposite situation. It had been commonly hunted in the earlier part of the nineteenth century, and by 1890 Macpherson wrote that *'its numbers have of late years greatly decreased'*, and advised sportsmen to preserve the species on account of its antiquity. The situation today is that it seems unlikely a wild population of the Pine Marten still exists in the county, despite those extensive forests of the alien conifers. Through this century there have been occasional specimens, reports of tracks and sightings – some by experienced wildlife observers – and in the winter of 1987-8 a Nature Conservancy Council survey claimed to find evidence of Pine Marten 'scats' in practically all the county's forest blocks. However, a repeat English Nature Survey in the autumn of 1993 failed to replicate any of these records. The animal has total protection under the Wildlife and Countryside Act, and has been included in English Nature's Species Recovery Programme, so this might not be the end of the story of the Marten's association with Cumbria.

The status of Macpherson's next pair of carnivores, the Weasel *(Mustela nivalis)* and the Stoat *(Mustela erminea)*, seems to have changed very little. Although they have been continuously trapped as a protection for game birds, yet they are both widely spread and probably common. The Stoat is seen more frequently, mainly because of its larger size, and the fact that some individuals turn conspicuously white in winter; but the tiny Weasel, pursuing its life among the small mammal runs in thick cover, is unnoticed by the casual observer.

Macpherson's three other county mustelids are very interesting, and to them we have now added a fourth.

The Polecat *(Mustela putorius)* was locally called the Foumart to distinguish it from the Sweet Mart (Pine Marten), a distinction which has some relevance for those of us who habitually apply our sense of smell to animal droppings! During the nineteenth century the species had been regularly hunted with dogs – sometimes, astonishingly, by night. At one time it had been common *'in almost every dale and on all the mosses'*, but by 1890 had become very scarce owing to the widespread use of the gin trap. Its extinction here probably occurred in the first decade of this century. This reflects the national picture in which by 1915 the Polecat was thought to have become restricted to Wales and the Marches. Reduced keepering through two world wars and the banning of the gin trap in 1958 allowed the Polecat to begin a recovery. Nationally it is spreading, but we should have had to wait a long time for its return to our county had it not been re-introduced. Animals from pure-bred Welsh stock have been released regularly over the last twenty years so that now Polecats are becoming quite numerous and breeding well in some parts of the Eden Valley. Recent road casualties have been noted from the Eden Valley, South Cumbria and west to Wigton and Bassenthwaite.

The question of the purity of this animal, genetically speaking, is complicated owing to its close relationship to the Ferret *(Mustela furo)*. There is more than a wisp of taxonomic mist about here, but the Ferret indeed may be derived from our Polecat, and any differences essentially the result of the Ferret's two thousand years of domestication. The Ferret is widely kept and frequently lost, and as it interbreeds freely with the Polecat there are numbers of this Ferret/Polecat cross in the wild population. Nevertheless, detailed work on skull measurements and DNA claim to have established the presence of the 'pure' Polecat in our Cumbrian fauna, and there it is now protected under Schedule VI of the Wildlife and Countryside Act.

The Otter *(Lutra lutra)* too provides a fascinating story of changing fortunes. Described by Macpherson as common from south to north of the county, he believed that, despite hunting with hounds, it was trapping alone which checked the Otter's increase. In 1905 Harry Britten, reporting on the 'Mammals of the Eden Valley', described it as plentiful in that valley – an area that had always been noted for the best hunting, *'especially about Wetheral, Armathwaite and Appleby'*. It is not thought that the Otter ever disappeared completely from the county, but a very serious decline in the late 1950s is generally attributed to the introduction of organo-chlorine pesticides – like dieldrin in sheep dip. At the top of its food chain, the Otter collected high levels of residual water pollutants which affected both its fertility and individual survival. Although restrictions were placed on dieldrin use in the early 1960s, Mitchell and Delap (1974) described the Otter as uncommon. Numbers have remained low for many years, but now there are some encouraging signs. Martin Twiss, then the Cumbria Wildlife Trust Otter Project Officer wrote to me in 1993, *'Cumbria appears to sustain a low but stable Otter population, although an encouraging expansion in both range and numbers in N.E. Cumbria has occurred over the past two to* three years, heralding a re-colonisation from the Scottish Borders'. He described the Otter as in reasonable numbers locally on the Border Esk, the Lyne, Irthing and lower and middle Eden. Recently Otters have moved back onto the Waver and Wampool in N.W. Cumbria after a gap of possibly two decades, and are now also present on the River Greta and at Bassenthwaite Lake. Occasional reports come from elsewhere within the National Park and S.W. Cumbria – yes, Otters *are* present and numbers *are* slowly recovering.

The Badger *(Meles meles)* is our great success story. By the end of the nineteenth century the animal had been the subject of a determined attempt at extermination over the previous two centuries. Indeed Macpherson believed it had become extinct as a free-living wild species by about 1830, and the odd records of Badgers in the county during the next sixty years all referred to escaped pets or animals deliberately introduced. It is difficult now not to think that on this Macpherson was mistaken, and by 1901, after several more reports, he was himself questioning whether the original stock had ever become entirely extinct. But Britten in 1905 said of the Badger, *'I have only one record of the occurrence of this animal. This was on the Skirwith Abbey Estate, near to Langwathby, where one was trapped by my father somewhere about 1888'*. Certainly the Badger was very rare, and it was not until after the First World War that numbers began to increase. It was at this time that Ritson Graham began to study the species – a study which led to the publication in 1946 of his paper, 'The Badger in Cumberland'. There he concluded that *'Though the present distribution of the Badger is fairly general throughout the county it is nevertheless patchy; there are parts of comparative density and others of relative scarcity'*. But he recognised that Badger numbers were increasing and the animal's range was expanding. This is the pattern that has continued. Despite road casualties and a continuing low level of persecution, the Badger is well distributed, and indeed numerous in the most suitable habitats. The Badger has had some legal protection for the last twenty years, and is now totally protected under its own Badgers Act (1973), to the extent that even its sett cannot be disturbed. Within the county there are regional Badger protection groups affiliated to a National Federation. The Badger's situation shows well the enormous power of public opinion as directed here to conservation.

Our last carnivore is one of which previous generations of British naturalists had had no experience. The Mink *(Mustela vison)* began to appear in the county as a feral animal first in the 1960s. It had originated as an escapee from fur farms, and it quickly began to exploit the riverine niche made available to it by the low numbers of Otters and the absence of the Polecat. Its numbers built up to high levels by 1980, despite a xenophobic campaign of trapping, and similarly xenophobic reports in local newspapers which warned children against swimming in the Eden for fear of attack! Trapping by the Environment Agency, and others, continues today. One trapper told me he had killed more than 500

Mink in the last twelve years. Despite this the Mink is now widely and possibly permanently established on all our major river systems. It is in low numbers because trapping and the animals natural territorial spacing have led with time to a more stable population.

Another particularly informative example of the change in attitude towards wildlife over the last century is seen in relation to the seals and cetaceans found around our coasts. In the nineteenth century, any living specimens were either shot or driven ashore and hacked to death, often to be rendered into oil. Today, beached specimens and injured or ailing individuals have had veterinary attention, and when possible have been returned to the sea.

Macpherson has few seal records, and identifies all but one – a vagrant Harp Seal *(Phoca groenlandica)* – as Common Seals. Seals have never been numerous this century, although the Grey Seal *(Halichoerus grypus)* has been known for many years, and usually now small numbers turn up regularly to winter in the Duddon estuary, while up to fourteen have been seen off the West Cumbrian coast and others have been recorded in the Solway. In recent years, four Common Seals *(Phoca vitulina)* have wintered on the Duddon, where they have been known to display the characteristic Common Seal behaviour of moving up river – swimming as far as Duddon Bridge. Almost certainly they have pupped on the estuary, as one very young pup has been found. This is a new breeding record for the county, and important as probably the only site on the west coast of England.

With the exception of the Porpoise, Macpherson also has few cetacean records. The Harbour Porpoise *(Phocoena phocoena)* and the Common Dolphin *(Delphinus delphis)* have both regularly been seen in the 1960s, sometimes in large numbers – up to thirty Dolphin and fifty Porpoise, but there has been a decline in the sightings of these smaller cetaceans over the last twenty-five years. This has been attributed to a decline in fish stocks, and an increase in pollution of the Irish Sea. Other cetacean records this century include Killer Whale *(Orcinus orca)*, Long-finned Pilot Whale *(Globicephala melas)* and Minke Whale *(Balaenoptera acutorostrata)*. Bottle-nosed Dolphin *(Tursiops truncatus)* have been regularly recorded over the last ten years off South Walney, and in the last five years in Morecambe Bay.

In any survey of Cumbrian mammals today we should expect to give some prominence to deer. But free-living deer in the last century were not so numerous. The Roe Deer *(Capreolus capreolus)* was restricted to the Borders, although on rare occasions some were known to cross the Eden and to wander up the valley of that river into the neighbourhood of Penrith. An odd animal near Rockcliffe and another near Stanwix were notable nineteenth century records. Martindale seems to be the only area where Macpherson thought wild Red Deer *(Cervus elaphus)* survived, noting some as moving to Place Fell in the winter months and stags as occasionally crossing Ullswater to Gowbarrow, while his account of Fallow Deer *(Dama dama)* is entirely based on those in parks, though it is to be presumed that a few at least existed ferally.

The status and distribution of deer have undergone dramatic changes this century; two world wars meant breached park walls and the growth of neglected scrub woodlands. Britten had remarked in 1905 on the occasional Roe stragglers in the Eden Valley, including several in Coombs Wood near Armathwaite. Interestingly, it was at Coombs Wood in 1953 that Peter Delap took the first photograph of a British Roe Deer. But in the post-Second World War years the Roe moved strongly into the county from the north. Others were known to be released into the wild at Hawkshead by the Curwens of Belle Isle. As they colonised new woodland their numbers grew to high levels, then crashed as individuals established territories and others moved on. Now the Roe is to be found in suitable habitat, generally distributed, and in particularly good numbers in the south of the county where complaints of its damage in gardens are numerous – especially from 'offcomers' who retire there to grow roses!

Red Deer similarly are well distributed. Stragglers from the three hundred or so in the Martindale Forest, and others, perhaps displaced by habitat degradation through sheep over-grazing but encouraged by coniferisation, have spread west to Haweswater and onto the Shap fells, and south towards Kendal. Some are based on the conifer blocks of Grizedale and Thirlmere. In the south of the county they are present in some limestone woodlands and on the mosses of Hay Bridge and Leighton. Even Middleton Fell near Kirkby Lonsdale retains a small population – probably derived from Rigmaden Park escapes. At the present the Red is secure, although, of course, dependent on human goodwill – and the price of venison.

Park escapes too have given us some feral Fallow Deer around Arnside and Yealand Conyers, and Chinese Muntjac *(Muntiacus reevesi)* in the late 1980s at Warcop, Kings Meaburn and Natland. Muntjac are still present in South Lakeland – as evidenced by a sighting at Brigsteer in 1995. It is unlikely now that many pure-bred feral Sika Deer *(Cervus nippon)* remain. This species hybridizes with Red Deer, and park escapes led to a confused situation in the Cartmel area in the early 1970s. It is thought that there were four or five in the Roudsea area about six years ago, and there may still be hybrids around Winster.

Mitchell and Delap mention the Feral Goat *(Capra hircus)* as a species present in the area of Christianbury Crags in 1974. This mammal is not recorded by Macpherson but was almost certainly present in the late nineteenth century. There is an intriguing, though uncertain, historical suggestion that this herd had been seen by Daniel Defoe when on his 'Tour through the Whole Island of Great Britain' in the early eighteenth century. The goats were certainly present in 1906 and numbers were as high as 86 in 1965. During the next decade, however, they declined, reaching about twenty in 1975 and perhaps ten in 1977. They are reputed locally to have been finally shot out in the late 1970s during a period of increased conifer planting of the Bewcastle fells. A new group of 15 animals consisting of 7 nannies, 7 kids and a yearling, from a Scottish source near Inverness, was introduced into Kentmere in August 1993.

The rodents have suffered mixed fortunes since 1890. Nationally, Red Squirrel *(Sciurus vulgaris)* numbers have fluctuated over the years, but generally there has been a decline since about 1930, probably owing to disease and loss of habitat. The national picture has been further complicated by the spread of the Grey Squirrel *(Sciurus carolinensis)* whose presence seems to have prevented successful re-colonisation of deciduous woodland by the Red. In Cumbria we are fortunate in still having good and widespread populations of Reds – both in conifer woodland and also in deciduous woods with some conifers, but that the Red Squirrel's long-term status is regarded as far from secure is evidenced by the recent implementation of the NPI Red Alert North West initiative to conserve the species.

New to the county this century is the Grey Squirrel. Introduced to Britain from the U.S.A. towards the end of the 19th century, it has slowly spread northwards since 1945. It is now widespread in the Arnside and Silverdale areas and in the last few years has been found at Ambleside, and has moved east and west to Sedbergh and Grizedale. There are now also a few isolated records in the north of the county. It is possible the two squirrel species might co-exist in coniferous woodland which favours the Reds, but the Grey seems better able to exploit deciduous woodland and over a number of years out-competes the Red with a greater breeding success. The modern trend to improve coniferous plantings aesthetically by fringing streams and edges with broadleaves may well favour the Grey. The Red Squirrel is now totally protected by law.

The same level of protection is afforded the Common Dormouse *(Muscardinus avellanarius)*, which has declined nationally through habitat fragmentation and loss. As a specialised arboreal species it is dependent on a diverse woodland habitat with a vigorous, unshaded, shrub layer producing berries and nuts. Commercial monocultural forestry is disastrous, and the species has been made a subject of English Nature's Species Recovery Programme in an attempt to consolidate the remaining populations. The Dormouse seems not to have been common in the county in the last century – Macpherson described its occurrence as sporadic *'in a few of the most densely planted portions of Lakeland as far north as Dalston'*, and unrecorded in the eastern parts of either Cumberland or Westmorland. Today it is still known to be present in two localities in the south of the county where English Nature has scheduled some areas of semi-natural woodland as Key Sites, and conservation of the very small populations is assisted with nest boxes and licensed monitoring.

Macpherson considered the Harvest Mouse *(Micromys minutus)* extremely rare in the county. He noted only two nineteenth century records and thought it possible that both arose through accidental introductions. Almost twenty years ago a single nest was identified at Temple Sowerby by a cereals inspector, and there have been a number of anecdotal reports of nests, especially around Langwathby relating to the same period. Experience has taught me to distrust all anecdotal reports relating to mammals, but this is a species which is most likely to be found in the northern half of the county and a survey may well reveal it to be more common than is at present supposed.

Macpherson's Long-tailed Field Mouse, now more usually called the Wood Mouse *(Apodemus sylvaticus)*, is regarded as probably the commonest wild mammal. It has the misfortune of attracting the attentions of at least thirteen known predators, but is very common everywhere, in gardens, scrub, woods and agricultural land. Similarly, the House Mouse *(Mus domesticus)*, despite continuous persecution, survives well in most situations, but especially in the locality of human habitations, farm buildings and food stores.

From Macpherson's references to the Black Rat *(Rattus rattus)*, now called the Ship Rat, the species seems to have lingered into the very late nineteenth century. Nationally, it is now mainly restricted to dockside areas. It is said that its habit of living mainly within buildings has made its numbers more susceptible to rodenticides than the Common Rat *(Rattus norvegicus)*. Formerly known as the Brown Rat, this latter species replaced the Black Rat in the county in the last century and is still very common in a great variety of habitats, but especially around all areas of human activity.

The Water Vole *(Arvicola terrestris)* too, was common last century and numerous for perhaps the first three quarters of the present one – especially in the lowland parts of the county. Mitchell and Delap in 1974 described it as *'well established in Lakeland'*. But there seems little doubt that this species has declined markedly in numbers in the last twenty years. Recent records are little more than a couple of 'plops' in the Eamont, where it had been plentiful in the 1970s, but it seems to have disappeared from sites in the south of the county – east of Urswick Tarn, and at Roudsea, where twenty years ago it was common. Some observers question whether Mink are responsible, and it seems likely that over-zealous management of river and ditch sides, and weed dredging, makes habitat unsuitable by exposing these mammals to excessive disturbance and predation. A recent national survey reported evidence of this species on the lowland stretches of several Cumbrian rivers, and a county survey is now urgently required.

Last century, Field Voles *(Microtus agrestis)* were generally common, and in fact Macpherson regarded them as increasing, owing to the destruction of their avian predators. They are still a very important prey species for Foxes, Kestrels and Barn Owls. Numbers may not be so high as formerly in meadows where silaging has replaced hay-timing, but they are still present commonly in the rough grassland of roadway embankments, and especially in young plantations – real beneficiaries of the commercial conifer cycle. They are always under-recorded in small mammal live-trapping surveys owing to the lack of incentive for these grassland specialists to enter traps baited with cereals.

Macpherson's comments on the Red Field Vole – our Bank Vole *(Clethrionomys glareolus)* – seem intriguing today. He describes the species as first ascertained to be resident in March 1887, and adds that it was in all probability *'extremely local'*. Britten (1905) says, *'I have only once seen this little animal in the Eden Valley'*, and adds, *'I believe it may be far from common'*. That this was not the case, however, is clear from a later (1909) footnote, where Britten adds, *'Further investigation proves that the Bank Vole is abundant in the Eden Valley as it is in many other parts of the county'*. The most likely explanation for the confusion is that the species was often mis-identified as the Field Vole. The Bank Vole is very common in deciduous woodland, scrub, hedgerows and gardens – particularly those with drystone walls which provide good overhead protection. Populations are, though, often subject to dramatic seasonal fluctuations. Unlike the Field Vole, the Bank Vole readily enters small mammal live traps.

Brown Hares *(Lepus europaeus)* Macpherson reported to be *'comparatively scarce'*, and quotes an observer who considered hares had been decreasing in Cumberland for the last sixty years. Today the distribution appears to be patchy with, for example, some areas of limestone grassland and arable habitats still having good numbers, while other populations seem reduced. Nationally Brown Hare numbers have declined over the last two decades and this is a species over which conservationists are becoming concerned.

The Mountain Hare *(Lepus timidus)* was introduced into southern Scotland about the middle of the nineteenth century. It is not clear when it first came into the county from the north, but it seems to have been present in the Borders at Spadeadam and Kershope in the 1950s and 60s, when it was recorded by Ritson Graham. Mitchell and Delap (1974) repeated these records and also referred to an introduction in the Coniston district in 1903, where Mountain Hares survived until the First World War. In the Borders it is thought that the animal was present until the late 1970s. By this date, however, extensive coniferisation of the area coupled with severe loss of heather through sheep over-grazing had rendered much of the habitat unsuitable for this species; it has not been recently reported.

The presence of the Rabbit *(Oryctolagus cuniculus)* needs no investigation. It was, and is, a very common species in the county. This continuing status did not always seem so assured. Its population went through a very dramatic decline in the middle of this century through the effects of the disease, Myxomatosis, which appeared in Britain in 1953 and for a time threatened the animal with extinction. It has since recovered, and populations have acquired some resistance, although the disease is now endemic and outbreaks usually occur in late summer every year, killing a proportion of the individuals. Numbers are further controlled by man, using a combination of gassing and more traditional methods. The recent occurrence of Rabbit Viral Haemorrhagic Disease in domestic Rabbits in the county gives cause for concern that the wild population could soon suffer from this disease.

This brief review of the species makes it evident how extensive generally is man's responsibility for the continuing survival of the county's mammals. Direct persecution, so common a century ago, has largely declined except in relation to those species, like the Rabbit, which are still regarded as pests. Of far greater concern are the economic forces which put pressure on habitats for development and exploitation. It is this habitat stress that is the greatest threat wildlife now has to survive, but on the evidence of the increasing knowledge and changing public attitudes in this century, we may on balance be optimistic about our mammals' success in the next hundred years.

Sources and references

Britten, H., (1909), Mammals of the Eden Valley. *Transactions of the Carlisle Natural History Society* **I**: 24–30.

Delap, P., (1970), Mammals. Pp. 176–193 in Hervey, G.A.K. & Barnes, J.A.G., *Natural History of the Lake District*. London: Warne.

Graham, R., (1933), The Roe Deer in Cumberland. *Transactions of the Carlisle Natural History Society* **V**: 104–116.

Graham, R., (1946), The Badger in Cumberland. Pp. 88–99 in *Lakeland Natural History*. Transactions of the Carlisle Natural History Society **VII**.

Graham, R., (Eds. Matthews, S. & Clarke, D.J.), (1993), *A Border Naturalist: the birds and wildlife of the Bewcastle Fells and Gilsland Moors 1930–1966*. Carlisle: Bookcase.

Johnston, B., (1933), Mammals in Carlisle. *Transactions of the Carlisle Natural History Society* **V**: 26–30.

Macpherson, H.A., (1892), *A Vertebrate Fauna of Lakeland*. Edinburgh: David Douglas.

Mitchell, W.R. & Delap, P., (1974), *Lakeland Mammals – a visitors handbook*. Clapham: Dalesman.

Norman, G., (1995), Bats and bridges in Cumbria. *Carlisle Naturalist* **3 (2)**: 36–37.

Strachan, R. & Jeffries, D.J., (1993), *The Water Vole, Arvicola terrestris, in Britain: its distribution and changing status*. London: Vincent Wildlife Trust.

Various Editors, (1973–1996), Mammal Reports in *Natural History in Cumbria* (retitled to *Birds in Cumbria* from 1977; *Birds and Wildlife in Cumbria* from

1996). Association of Natural History Societies in Cumbria (renamed Cumbria Naturalists' Union, 1990).

Webster, J., (1995), The return of the Polecat. *Carlisle Naturalist* **3 (1)**: 12–13.

WILDLIFE AND ITS CONSERVATION IN CUMBRIA

Derek Ratcliffe

I had the double good fortune to grow up in Cumbria. First, I was lucky in the early encouragement and inspiration that I received from the Carlisle Natural History Society and its leading members – especially Ernest Blezard – which helped to launch me on a career in natural history and nature conservation. Secondly, there can be few better places in the whole of Britain where an aspiring young naturalist might live and develop, and I count myself fortunate to have been so placed in my youth. The variety of country and wildlife within reach is remarkable, ranging from the great estuarine flats and salt marshes of the Solway; the sand dunes and sea-cliffs of the open coast; the farmland, woods and peat mosses of the lowlands; the rivers, lakes, valleys and rugged fells of Lakeland; and the rolling moorlands of the Pennines and Borders. And beyond Cumbria itself is the wonderfully varied country of southern Scotland, with the charms of Galloway pre-eminent.

The growth of conservation concern

One of the outcomes of observing the Cumbrian scene over half a century is the witnessing of change. Even left alone, nature never stands still, but it is the alterations contrived by the hand of humanity that force themselves especially upon attention, and become the focus of conservation concern. For, regrettably, much of this man-made change has not been to the benefit of wildlife and its habitat. 'Progress', in the form of land use developments, has left few parts of the region untouched by its impacts. Yet, as a counterweight to the damage and loss inflicted by human activity, there has been an enormous increase in enthusiasm for wild nature and concern for its conservation. The founding of the official Nature Conservancy in 1949, and the subsequent growth of the non-governmental conservation bodies (including national organisations such as the Royal Society for the Protection of Birds and Royal Society for Nature Conservation, and the local wildlife trusts) has led to programmes of site safeguard and wider environment measures that have given long-term protection against damaging change.

One of the earliest nature reserves in the country was established by the Carlisle Corporation at Kingmoor in 1913, largely through the influence of Linnaeus Hope and other local enthusiasts. Moor House in the Pennines was one of the first National Nature Reserves (NNRs) in 1952, and Ravenglass Dunes one of the first Local Nature Reserves, leading the way in protection of the most important wildlife areas in this country. And as an example of a 'wider environment' conservation issue, information from Cumbria also played a significant part in the presentation of evidence that persuaded Government to place increasing restrictions on the agricultural organochlorine insecticides which were causing such damage to some wildlife populations during the late 1950s and 1960s. These early developments paved the way to more comprehensive conservation programmes that followed.

The balance sheet of change

Geoffrey Halliday's paper has shown how human intervention has considerably enriched the Cumbrian flora, though most of the additions fall into one or other category of plant introductions, which do not have the same degree of importance for many people as the true native species. John Webster has referred to mammals such as the Badger and Roe Deer which have increased and spread markedly since H.A. Macpherson wrote his *Vertebrate Fauna of Lakeland*. Geoff Horne has charted the remarkable 'super-recovery' of the Peregrine population from its all-time low, and the increase of other bird predators such as Buzzard and Raven. For the butterflies and dragonflies, Geoff Naylor and David Clarke describe a variable pattern of increase and decline. Some of these changes are the result of either direct or indirect human influence, while others may be no more than the propensity to natural fluctuation in abundance and distribution. In the case of less well-known groups, such as the beetles reviewed by Roger Key, apparent increase may be more a reflection of increasing knowledge than real expansion in numbers or range.

Taking birds as the best known wildlife group, there are some positive changes to celebrate. We have seen how the last few decades have brought the return to Cumbria of the Golden Eagle, Hen Harrier and Goshawk as nesting birds; the rise of Peregrine population to unprecedented levels; and the local increase in Raven and Buzzard numbers. There has been a near explosion in numbers of Lesser Black-backed and Herring Gulls in coastal colonies, Kittiwakes have increased greatly at St Bees Head and Cormorants are now well established breeders. The district has also been colonised by Little Owl, Green Woodpecker, Siskin, Crossbill, Nuthatch, Collared Dove, Great Crested Grebe, Eider, Goosander and Red-breasted Merganser. The Ruff and Black-tailed Godwit have become included amongst the sporadic nesters. Feral colonies of Greylag and Canada Geese have established in various localities.

On the debit side of the equation we must include the decline of Merlin, Barn Owl, Long-eared Owl, Black Grouse, Red Grouse, Nightjar, Tree Pipit, Tree Sparrow, Yellow Wagtail, Corn Bunting, Curlew, Snipe, Golden Plover, Lapwing, Redshank and Little Tern, and the disappearance of the Corncrake. Despite the general recovery of the Sparrowhawk, this bird has not regained its former numbers in the Solway region, and both Raven and Buzzard have decreased in the Pennines. For some of these bird declines the causes are well understood, but for others they are obscure or uncertain.

Birds well illustrate the point that change has involved a balance sheet, in which there have been both gains and losses. Conservation concern has to allow for some fluctuations as a normal feature of the wildlife scene. It is when the losses show a consistent tendency to exceed the gains by a substantial margin that we have to worry. And it is in the wildlife habitats, especially those of natural or semi-natural type, where change tends to be a one-way process that gives rise to alarm. Most of my review will, accordingly, be devoted to a brief scan of Cumbrian habitats and the trends of change within these.

The coast

Beginning with the coast, I am impressed by how little the Solway area has altered in general appearance during the 50 or so years over which I have known it. The sprouting of the Chapelcross cooling towers on the Scottish skyline is the only obvious change. The great salt marshes have not suffered from further agricultural reclamation – though this was threatened at one point – and although the seaward edge of Burgh Marsh has steadily retreated through erosion, extensive new marsh has formed in recent years at the Drumburgh end. This is within the normal salt marsh cycle of accretion and erosion. Redshanks and perhaps Lapwings are no longer in their former numbers and Rockcliffe Marsh appears to be the only Cumbrian coastal nesting place of the Dunlin nowadays. Rockcliffe also has greatly increased numbers of the large gulls nesting, and has gained from protection as a Cumbria Wildlife Trust (CWT) Reserve. The Upper Solway (shared with Scotland) is one of the most important estuaries in Britain for its wintering wildfowl and waders, with numbers of Barnacle and Pink-footed Geese, Oystercatchers, Curlew, Dunlin, Knot, Bar-tailed Godwit and Golden Plover especially notable. Average peak winter counts during the five-winter period 1985/86 to 1989/90 were of nearly 110,000 waterfowl. This is probably the least affected of all major estuaries by urban-industrial development, though impending offshore oil exploration is a new threat, and a barrage scheme was put forward some thirty odd years ago.

The southern estuaries of Morecambe Bay (shared with Lancashire) and the Duddon are also largely unspoiled, with extensive systems of inter-tidal flats and fringing salt marshes. They are major bird haunts, and Morecambe Bay eclipses even the Solway as an area for wintering wildfowl and waders, with peak counts over the same five-year period exceeding 180,000 birds, and high numbers of Pink-footed Geese, Shelduck, Pintail, Wigeon, Oystercatcher, Knot, Dunlin, Curlew, Redshank, Bar-tailed Godwit and Lapwing. All three estuaries are internationally important for their birds. Morecambe Bay has been the subject of a feasibility study for a barrage to create freshwater supply, but the threat receded when Kielder Water was developed instead. The Duddon is more immediately threatened by proposals for an energy generation barrage across its entrance.

Seacliffs are among the more secure kinds of habitat and our most notable example, at St Bees Head, now has the added protection of RSPB ownership. It has the only breeding colony of auks, Kittiwakes and Fulmars on the mainland of north-west England. The main threat to the nesting birds is from oilspills and other marine pollution. This headland also has considerable interest in its plants and insects. The only other significant coastal cliff, at Humphrey Head (a CWT reserve), is a limestone promontory with great botanical interest, including four national rarities in Goldilocks Aster, Western Spiked Speedwell, Spotted Cat's-ear and Hoary Rock-rose.

The sand dunes which fringe the coast at intervals from Grune Point to Walney Island have changed little, and the most important of them are now protected as nature reserves. They support important populations of local plants such as Bloody Crane's-bill and Burnet Rose. Sand and shingle beaches remain important as plant habitats – where the Oysterplant has recently reappeared – and breeding places of Ringed Plovers and Oystercatchers. The collapse of the large Black-headed Gull colony at Drigg (Ravenglass) has been attributed to Fox predation, and the Little Tern has declined through recreational disturbance; but the Lesser Black-backed and Herring Gulls, Sandwich and Common Tern, and Eider colonies on Walney Island have increased through nature reserve protection. Brackish water habitats from the Solway to Morecambe Bay are important for Natterjack Toads, holding at least a quarter and possibly up to a half of the British population.

The Cumbrian coast has nevertheless had one of the least welcome of all developments, in the nuclear industry's complex at Sellafield, dispersing its particularly insidious pollutants into pathways that spread contamination both near and far, and with an invisible menace that many people find alarming.

Lowland grasslands and heaths

The largest and most serious habitat changes during recent times have probably been on farmland. While experiencing most of the trends of modern agriculture, Cumbria has suffered less from hedgerow removal than many more heavily arable counties. It has, however, been greatly affected by the grassland improvement that has converted most old, botanically diverse pastures and hay meadows to uniform species-poor swards. The ploughing, herbiciding, fertilising and reseeding with commercial grasses have destroyed the great bulk of the original communities of enclosed grasslands. Draining of fen and marsh often left areas of damp pasture with a good deal of wildlife interest, but this and the extensive rough grassland of the marginal land have now been extensively under-drained, and even ploughed, thereby reducing the habitats of moisture-loving plants and animals. The moorland edge has been driven back further by reclamation or improvement in many places, and many an untilled remnant in the lowlands has been brought into cultivation.

There has been a widespread loss of plants such as Globe-flower, Wood and Meadow Crane's-bill, Melancholy Thistle, Bird's-eye Primrose, Butterwort and many kinds of Orchid – Early-purple, Common Spotted, Northern and Early Marsh, Greater and Lesser Butterfly, Burnt, Small-white, Fragrant and Frog. The Wood Bitter-vetch has been a particular victim in the Pennine fell-foot meadows. Birds such as Snipe, Curlew, Redshank, Lapwing and Yellow Wagtail have declined in parallel, and the almost vanished Corncrake is a well-known casualty of the more intensive management and early cutting of hay crops.

Insects are much affected too. Cumbria was once a stronghold of the very local Marsh Fritillary butterfly, but this has lost most of its former haunts through pasture improvements that destroyed its larval food plant, the Devil's-bit Scabious. Although a meadow (Bucknill's Field) at Orton Woods was acquired as a reserve specifically to protect this insect, it is doubtfully large enough to hold a permanently viable population, since most of the adjoining meadows have been 'improved'. The last confirmed sighting of the Marsh Fritillary at Orton was in 1975. Recent loss of this butterfly at Finglandrigg NNR conpounds its threatened existence in the county. Others of the grassland butterflies are threatened by the impermanence of the herbaceous communities when left unmanaged. The once famous colony of the Small Blue at Cowran Cut, a deep railway cutting near Brampton, died out long ago when the natural processes of plant succession led to a smothering of its food plant, the Kidney Vetch, by dense growths of Bramble and other tall species. It persists elsewhere in the district, but its status is made tenuous by dependence on a food plant which grows especially in unstable or ephemeral communities.

These changes have enhanced the importance of the remnants of such herb-rich vegetation, especially on roadside and railway verges. It is particularly worrying, therefore, to find that even the road verges are under attack in some places. Some of the best examples of these communities now left in Cumbria are on the broad verges of clayey limestone drift soils north-west of Penrith. Here, at Hewer Hill, farmers have recently 'reclaimed' sections of verge by dumping topsoil and seeding it with commercial grasses. In places in this area there were stretches of an unusual and interesting grass-heath, where herb-rich grassland also had rather open growths of heather, and typical lime-loving plants such as Bird's-eye Primrose and Pepper-saxifrage were in company with others such as Common Wintergreen and Petty Whin which usually belong to acid soils. The best example, at Johnby Moor, was largely lost to afforestation. Elsewhere on the New Red Sandstone of the Carlisle district were good stands of more typical lowland acidic heath of southern type, with Ling, Bell Heather and Cross-leaved Heath. Extensive areas on the Eden valley sandstone at King Harry, Ainstable, Blaze Fell, Baronwood Park and elsewhere have been ploughed and converted to grassland, though good areas on Wan and Lazonby Fells survive. Part of a good example with Western Gorse at Walby Moor was reclaimed to arable, but an especially interesting type with abundance of the southern Dwarf Gorse lies within the NNR at Finglandrigg.

Lowland bogs and fens

The lowland 'peat mosses' are mainly raised bogs on the coastal plains, but with a few basin and valley bogs in undulating glacial drift country, especially east of Carlisle, and others intermediate between raised and blanket bog around Hethersgill and on Denton Fell. The largest raised bogs are those of the Solway, but there are others around the edges of the Duddon estuary and Morecambe Bay. In the north, Solway Moss has largely been lost to commercial peat extraction, and a large part of Wedholme Flow has been cut over and spoiled, but the peat domes at Glasson, Bowness and Drumburgh Mosses are still largely intact. Repeated fires over the last 40 years have caused a large reduction in *Sphagnum* cover (responsible for active growth of the bog surfaces) on most of these mosses, though the flora survives almost without loss. Glasson Moss is now a NNR where the vegetation is recovering under careful management, and Drumburgh Moss is a CWT Reserve, but fire remains a serious threat to all these mosses. Of the southern raised bogs, Foulshaw Moss is almost obliterated by conifer forest, and the Marsh Gentian which grew there appears to be extinct, but most of the other areas appear to have survived.

Most of the basin and valley bogs also remain, as at Moorthwaite and Cumwhitton, though some have suffered the effects of nutrient enrichment through inflow of fertiliser from surrounding farmland. Spontaneous pine colonisation is also a problem on some sites. Of the intermediate bogs, Bolton Fell has been extensively worked for horticultural moss-litter during recent years, and Denton Fell is largely afforested, but Walton-Broomhill Mosses remain fairly intact.

The peat mosses remain important sites for plants such as Bog-rosemary, Bog Asphodel, Cranberry and Great Sundew, which are still abundant. The rare Bog Bilberry grows in several bogs to the east of Carlisle, and is in spectacular abundance in one of them. The fringing birch and pinewoods are also quite valuable habitats in their own right. Although the bird interest has declined, with loss of the Nightjar, Merlin and breeding gull colonies, the mosses remain a major haunt of the Adder and local insects such as the Large Heath Butterfly and Emperor Moth.

Cumbria has not been particularly well endowed with lowland fens, though these wetlands may have been drained out of existence in ancient times, as the early human settlers strove to create farmland. We know that important remnants at Cardew Mires near Dalston and Tarn Wadling by High Hesket were drained during the 19th century. Sandford Mire at Warcop was lost some time after this and remnants survived until fairly recently. One of our best remaining examples, at Newton Reigny Moss, near Penrith, has dried out substantially since the cutting of a drain some 50 years ago, losing some of its important plants such as Marsh Helleborine, Bird's-eye Primrose, Greater and Lesser Bladderworts and Bog-sedge. Biglands Bog, with its remarkable nucleus of acidic bog surrounded by fen, has been adversely affected by the inflow of sewage from adjoining farms. Sunbiggin Tarn, with perhaps the best example of marginal calcareous fen, is evidently deteriorating through the cumulative effects of many years' manuring from a large Black-headed Gull colony. Some of the lakes have fragments of fringing fen, of which Esthwaite North Fen is probably the most important. There are also many examples of mires with some features of acidic bogs, known as 'poor fens', with Tarn Moss NNR at Troutbeck (Matterdale), Halsenna Moss NNR near Seascale, and Cliburn Moss near Penrith among the best. Most of the surviving sites named above are protected in some degree by reserve status, but their topographic position inevitably makes them vulnerable to catchment influences over which there is little, if any, control.

Woodland

Woodland is one of the most important habitats for wildlife, but especially the ancient semi-natural types, composed of native trees which have occupied the site continuously since the post-glacial spread of forest. Some of the wild-woods on or beside the lowland peat mosses are little altered (such as those at Orton Moss and Finglandrigg near Carlisle), but some fine broadleaved woods have been ruined since the last war by felling and replanting with conifers (eg Coombs Wood at Armathwaite, Park Wood at Brocklebank and Walton Wood near Brampton). Plantations have some wildlife interest, and stands of old Scots Pines have been a noted habitat of Creeping Lady's-tresses in the district, but on the whole they have far less variety of both plants and animals than ancient native woods. Recent planting has been mainly with exotics – Sitka Spruce, Douglas Fir and Larch which cast such deep shade and/or have such heavy needle fall that the ground flora becomes greatly reduced. There has been some grubbing out of old woodland to create more farmland, *eg* part of Thurstonfield Woods. Few woods are now managed as coppice, so that woodland plants and associated insects which depend on open phases of the forest cycle have tended to decline.

The rocky fellside woods of Oak or Oak, Ash, Birch, Wych Elm and Hazel in Lakeland are of special interest for their communities of ferns, mosses, liverworts and lichens which flourish under the humid climate, and particularly in the shady waterfall ravines. This is the richest area in England for these plants and, while Borrowdale (Derwentwater) is outstanding, the Ullswater, Buttermere, Thirlmere and Eskdale valleys also have important woods and glens. Some of the finest woods, especially in Borrowdale, are in the care of the

National Trust, who well understand the need to maintain tree cover, of broadleaves, to protect the moisture-loving 'lower' plants, in managing these sites. In Lakeland generally, it seems to have been increasingly accepted by woodland owners and managers that existing broadleaved woods should be replanted with native species rather than conifers.

One of the tragedies of the conservation scene is the way in which Dutch Elm disease has eventually rampaged through almost the entire native Wych Elm population of Cumbria. Apart from the loss of the intrinsic value of this once abundant tree, its base-rich bark is one of the most important habitats for bryophytes and lichens. Atmospheric pollution by acidifying gases has also reduced the abundance of many lichens and mosses on trees, though there are indications that the process may be reversible, if pollution levels can be reduced sufficiently.

On the Carboniferous limestone and in places on other basic rocks such as the Silurian slates there is mixed broadleaved woodland with abundant Ash. Fine examples occur at Roudsea Wood NNR, Witherslack, Helbeck-Swindale (Brough), Argill and Smardale, the last two being CWT Reserves. These sites have rock habitats which enhance their botanical diversity, and more fragmentary examples on ravine sides (eg Irthing Gorge at Gilsland, Podgill at Kirkby Stephen) have an especially rich flora, including the bryophytes.

A characteristic feature of the Lake fells is the local occurrence of Juniper scrub, which in southern areas often has abundant Yew. These Juniper scrubs are retreating slowly, sometimes as the result of deliberate burning but also from the death of aged trees, and because heavy grazing suppresses the seedlings which would produce their regeneration. Lack of regeneration of trees in hill woods open to sheep is a general problem. Although the trees are mostly long-lived, there must come a time when such woods are effectively moribund, unless action is taken to promote the establishment of new young trees, especially through fencing of compartments against stock for a critical period.

Lakes and rivers

The lakes and lesser tarns are a major habitat of the district, and one of the best known, through the studies of the Freshwater Biological Association based on Windermere. They show a range of water chemistry and associated plant and animal communities, though most fall in the band having from low to moderate nutrient content. Thirlmere and Haweswater have lost much of their original ecological interest through the raising of their water levels, and the creation of unsightly fluctuating shoreline draw-down zones. There is evidence for acidification of some small Cumbrian tarns with unbuffered, nutrient-poor waters, but the main threat to the larger lakes appears to be from nitrogen and phosphate enrichment through the inflow of sewage and agricultural fertiliser.

There is concern that symptoms of such effects are appearing in Bassenthwaite Lake which, with Derwentwater, is now the only locality for the Vendace. Bassenthwaite Lake is the most recently declared of the Cumbrian NNRs, and an important addition to the series. Boating disturbance on lakes can adversely affect birds, but is at a minimum during winter when bird numbers are highest. Windermere is the most important lake for birds, and some of the nutrient-poor lakes have rather little ornithological interest. The artificial tarn of Monkhill Lough west of Carlisle was a good bird haunt, but was drained in recent years by removal of the dam.

The rivers of Cumbria are of considerable interest as a wildlife haunt, and the largest, the Eden, is one of the major English rivers least affected by pollution or gross physical modification. There is again a wide range of water chemistry, from the acid peaty headstreams draining the eastern and northern moorlands, to the hard-water gills of the limestone areas. Acidification effects have been found in rivers draining base-poor catchments, such as the Esk and Duddon. Many streams of the upper catchments run through ravines below hanging valleys, while the lower courses of the main rivers draining to the Solway have cut rocky gorges through the New Red Sandstone or the Carboniferous series. Both the upland ravines such as Lodore, Dalegarth, Launchy Gill and Rydal Beck, and lowland glens such as the Nunnery, Baronwood, Gelt Woods and the rocky sections of the Lyne and Irthing are important for their moss and liverwort assemblages. The rivers and lakes together form the haunt of the Otter, which has maintained its numbers more successfully in Cumbria than in many other parts of England.

Mountains and moorlands

The fell country forms the largest area of undeveloped land in the district, though it is mostly far from natural. Most of it was long ago stripped of the original tree cover which ascended the hill slopes to at least 500 metres, or even higher in places. Only a fragmentary patchwork of native woodland remains, and is almost everywhere below its climatic limit. Most of the higher open fells pass below into the enclosed fields of the upper farms, though there is often a broad transition zone of marginal land. Even the hill meadows have nearly all suffered the improvements of modern agriculture, and in the spring many of them show an unnatural hue of vivid emerald green which bespeaks the large addition of NPK fertilisers. Only a few good examples remain, one being the Gowk Bank NNR on the upper Irthing. The enthusiasm for moor-gripping (draining) has passed, since the Ministry of Agriculture withdrew its grant-aid for this, but not before a good deal of damage was done to small, botanically rich upland marshes and flushes. One drain narrowly missed a flush with the rare Marsh Saxifrage on the moorlands behind Crossfell.

Since deforestation, the most significant change has been the widespread conversion of the heather moorland (that usually replaced the forests) to acidic grasslands, often infested with bracken on dry ground, under the influence of heavy grazing and fire. The change has evidently proceeded over the centuries of management of the hills as grazing range for cattle, goats and latterly, sheep, so that most of the Lake fells and the Eden valley side of the Pennines are green and grassy. Heather ground is found mainly on rocky terrain, on the poorest rocks and soils (such as Skiddaw Slate) and on blanket bogs, where it is mixed with cotton grass. Management for Red Grouse has aimed at conserving heather, as the main food plant of this bird, but where sheep have gained the upper hand and the heather has retreated, grouse moors have been abandoned, as at Faulds Brow above Caldbeck. In many places, Bilberry becomes dominant as an intermediate stage in the replacement of heather moor by grassland.

Management agreements between the National Trust, English Nature and the local farmers on the Armboth Fells, involving paying for a reduction in sheep numbers, are trying to reverse the loss of heather. It will be even more important to extend this approach to Skiddaw Forest, which has greater bird interest, and where the contrast between management as sheepwalk and grouse moor is strikingly shown. Heather and associated dwarf shrubs provide Red Grouse with food and Merlins with nesting habitat, while the latter also takes day-flying moths (*eg* Emperor, Fox and Northern Eggar) which depend on these plants. Some of the dwarf shrubs themselves have become rare, even on heather ground, where they regenerate less freely than Ling and Bilberry after fire, *eg* the southern Petty Whin and the northern Bearberry.

The sub-alpine zone of dwarf shrub heaths and grasslands, lying below the natural tree-line, is the main feeding area for upland birds and other animals. It has in the Borders and even some outlying parts of Lakeland become much subject to afforestation with exotic conifers, which has transformed the land and its wildlife – replacing moorland with forest on a large scale, draining all but the wettest bogs and leaving open ground only on high watersheds, around rock outcrops and along rides or roadsides. During the first years after planting there is often good wildlife habitat. The existing vegetation grows tall and herbs flower freely. Field Voles and other small rodents multiply and attract Short-eared Owls and Kestrels, even Hen Harriers, while the increased cover encourages Black Grouse and small birds such as Whinchats and Grasshopper Warblers. But most of the waders soon disappear and when the tree canopy closes, at 10-15 years, the moorland bird community with Raven, Red Grouse, Curlew, Golden Plover, Dunlin, Ring Ouzel and Wheatear is replaced by a woodland one with Wood Pigeon, Tawny Owl, Song Thrush, Chaffinch, Dunnock, Robin and Goldcrest. A few rarer species are attracted to the new forests, such as the Crossbill, Siskin and re-introduced Goshawk, and they have encouraged the increase and spread of the Roe Deer on the Borders. Afforestation of a large part of Greystoke Park appears to have been partly responsible for the decline of the Buzzard population there, from 12 pairs to 2, but the loss of Rabbits through Myxomatosis was evidently a contributory cause. The greatest impact of the new forests is on the plantlife, which all but disappears under the thicket stage of the plantations, and becomes confined to the limited areas of unplanted ground. In lowland and sheltered situations where the forests can be thinned, there may be re-development of a limited flora, especially with ferns and mosses. High-lying forests are likely to stay unthinned, because of windthrow hazards, and these remain extremely poor botanical habitats. After clear-felling, open ground vegetation re-establishes, but consists mainly of common species, and lasts only until the re-planted trees close to form forest again. There is usually a net loss of flora through afforestation, with species of wet ground declining particularly. On the Bewcastle-Gilsland moors, plants such as Bog-rosemary, Cranberry, Broad-leaved Cottongrass, Few-flowered Sedge and Tall Bog-sedge have decreased through afforestation, and the only known colony of Mossy Saxifrage has been lost.

Now that so much of the Bewcastle-Gilsland moors have been lost to conifer forests, the Pennines are the only extensive area of moorland habitats in Cumbria. The RSPB's Geltsdale reserve has a varied complex of habitats – heather moorland and blanket bog, but also hill woodland, extensive marginal land, lake (Tindale Tarn) and stream. Its bird populations are diverse and important. Moor House NNR is higher, and has large areas of blanket bog and acidic grassland, but also good limestone habitats, and rises to stony fell-field on the watershed of the Crossfell range. There are also numerous occurrences of overgrown lead-mine spoil, with a distinctive flora. Grouse are still widely preserved on the Pennine moors, and it is here where persecution of predators is still a problem in places. Hen Harriers are certainly unwelcome and there are signs that tolerance recently extended to the Peregrine may be fading again. It is surely no coincidence that the Raven has almost disappeared as a nesting bird from the Pennines, though the problem may be mainly poison put down for Foxes and Crows. The Buzzard has also declined seriously along the Pennine scarp of the Eden valley, though perhaps more through the decimation of the rabbit population from Myxomatosis in this area.

The Lake District is a craggier region than the Pennines, so that plants which need protection from grazing and birds dependent on cliffs for nesting are in greater abundance here. Many alpines need lime in the soil and are rare because there is so little calcareous rock at high elevations in Lakeland: the Helvellyn range is their best area. Yet the most famous mountain plant, Alpine Catchfly, grows on acidic Skiddaw Slate at Hobcarton Crag. Some of the mountain plants may have gained from heavy exploitation of the uplands as grazing range, such as Alpine Lady's-mantle in some close-cropped grasslands and Parsley Fern on the extensive screes exposed by deforestation. Others have been reduced to minute populations, first by post-glacial climatic changes and woodland advance, then by land use effects. Plants such as Mountain Avens, Downy Willow, Shrubby Cinquefoil and Bearberry appear to owe their highly relict status to these processes but, once reduced in this way, some species have been made rarer still by plant collecting during the last 150 years. Rare ferns such as Oblong Woodsia and Holly-fern have been almost exterminated by collecting, but Alpine Cat's-tail, Alpine Saxifrage and Alpine Catchfly have also had their small populations reduced.

Limestone is usually the key to botanical richness, and the old county of Westmorland has some of the most extensive exposures in Britain, ranging from low-lying examples around Kendal and Orton, to the high Pennines, where the amount of this rock compensates for the lack of crags, as good habitat for mountain plants. One of the most distinctive features is the limestone pavements – tabular, flat or gently sloping and heavily fissured exposures, of which there are fine examples at Hutton Roof Crag-Farleton Knott and Great Asby Scar, with many rare or local plants, such as Dark-red Helleborine and Rigid Buckler-fern. One of the most disastrous activities from a conservation viewpoint has been the removal of the weathered surfaces of these pavements for rockery stone, which has damaged some irreparably. NNRs over part of Hutton Roof Crag and Great Asby Scar have protected some, and most of the other important examples are covered by Limestone Pavement Orders, but some surreptitious removal still goes on. Limestone cliffs and screes are still important, as at Scout and Whitbarrow Scars, but even these are not safe, and there have been problems with climbing disturbance and 'gardening' of crags in places.

The high limestone pastures of the Pennines are the nearest to alpine meadows that we have in England, and those on Little Fell above Brough are a fine example, with abundant Alpine Forget-me-not, Mossy Saxifrage and Spring Sandwort, and a scatter of Spring Gentian. The increased grazing pressure from growth in sheep numbers is a problem here, since there is a limit to what any plants except some grasses can withstand. Even under freehold ownership, the Moor House NNR is bedevilled by common grazing rights, which prevent control of the grazing regime. The effects of grazing have extended even to the high fell-field summits of Lakeland and the Pennines, the haunt of Dwarf Willow, Stiff Sedge and Dotterel. These places once had extensive carpets of Woolly Fringe Moss, but even this has declined and almost disappeared from the summit heaths of the Lake fells, and the Dwarf Willow has gone from a few former stations such as Crossfell. There is plenty of the moss left on stable block screes and it looks as though the grazing, manuring and treading of sheep (possibly assisted also by human trampling) have been responsible for its replacement by grasses. The failure of the Dotterel to recover its previous status as a nesting bird in Cumbria may be related to a decline in its food supply on the high tops, possibly involving such effects but also acidification.

Other kinds of change

Some forms of human impact are not specific to particular habitats. Cumbria has had its share of intrusions by the Ministry of Defence and, although two of the firing ranges are in the uplands, at Warcop and Spadeadam, that at Eskmeals is located amongst coastal dunes. The current enthusiasm for wind farms as a source of clean energy is producing applications for a variety of situations, from the coast to the hills. Both defence lands and wind farms are claimed to bring indirect benefits for wildlife, but they are both highly destructive to that wilderness character of undeveloped areas that so many naturalists crave. Cell-phone relay towers are yet another form of escalating visual intrusiveness. Recreational use of the countryside rightly expands in many different habitats, but has a negative aspect when it becomes too heavy or demanding. Atmospheric pollution, including acid rain, affects a wide range of habitats, though especially those where substrates and water are already somewhat acidic and base-deficient.

Assessment of conservation achievements and prospects

Cumbria is still clearly a wonderful area for the natural historian, with an exceptionally varied range of habitats and a correspondingly rich flora and fauna. Yet this biological interest has suffered significant reduction during the past half century, especially in its botanical aspects, through various forms of human impact; and there is a continuing attrition of wildlife value that must be regarded as a matter of great conservation concern. It is true that human agency has produced some additions and compensations, in habitats such as abandoned gravel workings, new tarns, old quarries, mine spoil, tree plantations, road and railway verges. There are also the numerous introduced plant species established in the wild, and a smaller number of animals gone feral. But most naturalists feel that the overall picture is of appreciable net loss, which calls for a conservation programme as a defence against this trend.

Spearheading such a programme is the safeguarding of the best wildlife areas under some kind of nature reserve designation. While it is clear that nature reserve status affords the strongest protection to wildlife, by controlling land use and the ecological factors which impinge on plants and animals, there are limits even then to what it can achieve. The freehold NNR of Moor House has to contend not only with common grazing, but also with barytes mining (mineral rights held by a mining company), a hill-top radar installation with access road, resultant winter skiing, motor bikes, mountain bikes and the Pennine Way. Whether it will be proof against pressure to re-open the grouse shooting remains to be seen. Special site protection can nevertheless be an effective means of safeguarding the very best and most vulnerable examples of habitats, and the biggest concentrations of notable species – especially colonial birds. The existing nature reserve series is thus a crucial cornerstone of the whole conservation programme, though the Appendix list shows that many of the Cumbrian reserves are small (some are minute) and vulnerable to adverse change on surrounding land. However secure the nature reserve designation might be made, the over-riding problem is that funds are far too short for it to be applied to more than a tiny fraction of the total area deserving protection. Site of Special Scientific Interest is a half-hearted mechanism, designed to achieve consultation over development proposals, and with a highly uncertain prospect of mitigating adverse effects, whether under planning law or the agriculture-forestry provisions. The SSSI record is a catalogue of continuing erosion of special wildlife interest through damaging operations in land use on these important sites.

There remains the much larger unprotected generality of the wider countryside, with the bulk of habitats and the populations of widely dispersed species. Some damaging environmental influences are also so pervasive that nature reserves and other designated special sites are as vulnerable to them as anywhere else – the fall-out of acid deposition, for example. We thus have to rely on the even more uncertain and unsatisfactory approach of trying to influence landowners and users, developers, planners and politicians – both central and local – who determine environmental policy and practice in all directions. The species protection laws, under the Wildlife and Countryside Act 1981, go only so far in safeguarding wildlife against some of the more direct and deliberate interferences or attacks, and have no power to restrain those who create incidental damage through legitimate development or other land use activity outside the SSSIs. The Planning Board of the Lake District National Park takes as sympathetic a view of nature conservation as it is able, but has only limited control over farming and forestry matters. One hopes, too, that the Area of Outstanding Natural Beauty designation for the North Pennines will take account of nature conservation requirements. The National Trust, as the largest single landowner in the district, is also responsive to conservation needs in the management of its properties, but has to balance these against the priorities of its tenant farmers and the need to earn revenue.

We need to persuade central Government of the need for overall policies which take adequate account of nature conservation needs – recognising that it is uphill work in the case of Departments such as Transport and Energy. The Ministry of Agriculture is improving slowly, and the provisions for reducing farm output, such as Environmentally Sensitive Areas and Setaside, give considerable opportunity for taking account of wildlife needs. The Forestry Commission affirms its commitment to conservation, including the safeguard of all SSSIs in its ownership, but the biggest problem with forestry is in preventing the further spread of new blanket coniferous afforestation. The Department of the Environment does not see its role as that of environmental champion, but rather one of reconciling conflict between different factions, taking a planners' view and defusing situations embarrassing to political masters. We wait with great concern and some anxiety to see whether its reconstructed nature conservation arm, English Nature, will show firmness and initiative in the exercise of its duties. Its continuation of the programme for designation of National Nature Reserves is greatly to be welcomed. Among the other statutory bodies, the National Rivers Authority [now the Environment Agency] has considerable influence on the wildlife conservation scene.

Local politicians will always tend to be persuaded by the job opportunities and wealth creation of development, so we have to try to direct development to the places where it will cause least damage. This means trying to influence the planners in the first place, since they are able to take a long-term, strategic view of development. Then there are the individual interests in development, who all have all to be approached directly: in farming, forestry, energy generation, minerals extraction, urban-industrial development, transportation, recreation and defence.

While much of this work of persuasion is the job of full-time professional conservationists, they will be severely limited without the support of the concerned naturalists and sympathetic public. Not only is help needed in managing the local reserves and in fund-raising, but above all in moral support and political pressure, through lobbying and campaigning. The question of how much wild nature will survive in Cumbria in the years ahead will be largely a measure of how much people care about it and are prepared to support and, if necessary, fight for it. This applies especially to those wider concerns which are perhaps the most alarming of all: the various aspects of atmospheric pollution that have given us acid deposition and now dangle over us the Damocles sword of global warming and sea level rise.

Sources and further reading

Blezard, D., (1946), Creeping Lady's Tresses in Cumberland. Pp. 71-4 in *Lakeland Natural History*. Transactions of the Carlisle Natural History Society **VII**.

Blezard, E., Garnett, M., Graham, R. and Johnston, T.L., (1943), *The Birds of Lakeland*. Transactions of the Carlisle Natural History Society **VI**: 1-170.

Cumbria Naturalists' Union, (1993), *Birds in Cumbria: a County Natural History Report*. **Jan-Dec 1992**. Carlisle: CNU.

Cumbria Wildlife Trust, *Reserves Handbook*. Windermere: CWT.

Hodgson, W., (1898), *Flora of Cumberland*. Carlisle: Meals.

Macpherson, H.A., (1892), *A Vertebrate Fauna of Lakeland*. Edinburgh: D. Douglas.

Nature Conservancy Council, (1984), *Nature Conservation in Great Britain*. Shrewsbury: NCC.

Nature Conservancy Council, (1984), *Nature Conservation and Afforestation in Britain.* Peterborough: NCC.

Nature Conservancy Council, (1991), *17th Annual Report* 1 April 1990-31 March 1991. Peterborough: NCC.

Pennington, W., (1973), *The Lake District.* London: Collins (New Naturalist series, no.53).

Pritchard, D.E., Housdon, S.D., Mudge, G.P., Galbraith, C.A. and Pienkowski, M.W. (Editors), (1992), *Important Bird Areas in the UK, including the Channel Islands and Isle of Man.* Sandy: Royal Society for the Protection of Birds.

Ratcliffe, D.A., (1960), The Mountain Flora of Lakeland. *Proceedings of the Botanical Society of the British Isles* **4**: 1-25.

Ratcliffe, D.A., (Ed.), (1977), *A Nature Conservation Review.* 2 vols. Cambridge: Cambridge University Press.

Stokoe, R., (1952), The birds of the Lake Counties. Pp. 13-112 in *The Birds of the Lake Counties.* Transactions of the Carlisle Natural History Society **10**.

Wilson, A., (1938), *The Flora of Westmorland.* Arbroath: Buncle.

Editors' note: English and scientific names of plants follow *New Flora of the British Isles,* 1991, C.A. Stace (C.U.P.)

APPENDIX: NATIONAL NATURE RESERVES IN CUMBRIA

1. National Nature Reserves

Asby Scar	307ha	NY 657097
Bassenthwaite Lake	523ha	NY 215295
Blelham Bog	2ha	NY 363005
Clawthorpe Fell	14ha	SD 537787, 535784
Finglandrigg Woods	65ha	NY 275568
Gowk Bank	15ha	NY 679739
Halsenna Moor	24ha	NY 066007
Moor House	3894ha	NY 735325
North Fen	2ha	SD 357977
North Walney	148ha	SD 170713
Park Wood	15ha	SD 565777
Roudsea Woods and Mosses	423 ha	SD 333820
Rusland Moss	24ha	SD 335886

Wildlife and its conservation in Cumbria

South Solway Mosses	560ha	
Glasson Moss		NY 235604
Bowness Common		NY 205601
Wedholme Flow		NY 220530
Tarn Moss	16ha	NY 400275
Thornhill Moss	12ha	NY 174486

2. **Royal Society for the Protection of Birds Reserves**

Campfield Marsh	310ha	NY 190610
Geltsdale	5083ha	NY 600530
Haweswater	9503ha	NY 500130
Hodbarrow	105ha	SD 175784
St Bees Head	22ha	NY 940145

3. **Cumbria Wildlife Trust Reserves**

Allan Wilson Memorial Reserve	1.8ha	NY 457526
Argill Woods	7.4ha	NY 844141
Ash Landing	2.3ha	SD 386952
Barkbooth Lot	12ha	SD 415906
Beachwood	0.8ha	SD 452786
Blawith & Brown Robin	26ha	SD 411791
Boathouse Field	0.6ha	NY 253231
Bowness-on-Solway	6.4ha	NY 207618
Bucknill's Field	0.6ha	NY 338543
Burns Beck Moss	15.1ha	SD 596881
Christcliff Duck Field	0.4ha	NY 185007
Clints Quarry	9.2ha	NY 009124
Causeway End Heronry	0.4ha	closed reserve
Dorothy Farrer's Spring Wood	2.6ha	SD 480983
Drumburgh Moss	119ha	NY 256590
Dubbs Moss	7.2ha	NY 104288
Enid Maples	2.8ha	SD 526897
Eskmeals Dunes	67ha	SD 087944
Foulney Island	11ha	SD 243655
Goldrill	1.1ha	closed reserve

Grubbin's Wood	7.2ha	SD 445780
Hale Moss	3ha	SD 510776
Humphrey Head	23ha	SD 391738
Hutton Roof Crags	99ha	SD 541777
Ivy Crag Wood	1.8ha	NY 245265
Juniper Scar	0.4ha	closed reserve
Latterbarrow	4ha	SD 440828
Meathop Moss	64ha	SD 445820
Newton Reigny Moss	0.3ha	NY 477312
Next Ness	1ha	SD 302787
North Walney NNR	*650ha	SD 170720
Rockcliffe Marsh	1120ha	NY 340637
Smardale Gill	40ha	NY 738083
South Walney	487ha	SD 215620
Tarn Sike	2.6ha	NY 665076
Waitby Greenriggs	4.4ha	NY 757086
Whitbarrow, Hervey Reserve	100ha	SD 442871
Willow Pond	0.2ha	closed reserve
Wreay Woods	19ha	NY 444500

*includes NNR and intertidal area

4. National Trust Reserve

Sandscale Haws	282ha	SD 190749

(The NT also manages substantial areas of land in the county, particularly within the Lake District National Park. This land holding incorporates all or part of 45 SSSIs and 2 NNRs. Nature conservation is an increasing priority of management plans on many properties.)

5. Other Reserves

Harrington Reservoir (Allerdale)	7ha	NX 994258
Kingmoor (Carlisle City Council)	44ha	NY 388580
Kingmoor Sidings (Carlisle City)	9.15ha	NY 387575
Ravenglass Dunes (LDNP)	383ha	SD 070960
Siddick Pond LNR (Allerdale)	19ha	NY 001305

CARLISLE NATURAL HISTORY SOCIETY

– the first hundred years

Stephen Hewitt

On 30 November 1893, four Carlisle teenagers with a mutual interest in butterflies and moths organised a meeting to form an entomological society. Thus was born the Carlisle Entomological Society, notice of which appeared in *The Entomologist* for January 1894, stating that 12 members had been enrolled on that first evening. One of the founding members of the Society was the 18-year old Frank Henry Day who later published an account of the early history of the Society in Volume V of these *Transactions* (Day, 1933).

That first meeting was held at the house of Mr Christopher Eales. However, Eales was an elderly man and future meetings planned at this venue could not take place due to his failing health. At this point the embryo Society might well have foundered but for the intervention of the Reverend Hugh Alexander Macpherson. Although an ornithologist rather than an entomologist, Macpherson was keen to support and encourage all branches of natural history study and he invited the young entomologists to a meeting at his house. As a result of this meeting it was decided to widen the scope of the Society to cover all aspects of natural history and attract new members interested in ornithology and botany. Macpherson was elected the first president of the renamed Carlisle Entomological and Natural History Society at its inaugural meeting on 1 February 1894.

Macpherson's reputation and energy did much to further the development of the new Society. Born in Calcutta in 1858, he was educated in England and came to Carlisle in 1882 as Curate of St James' Church (Hope, 1912). A passionate ornithologist, he set about studying the fauna of the county with great enthusiasm. He contributed many notes and papers to *The Zoologist* and other journals, as well as publishing in the *Transactions of the Cumberland and Westmorland Association for the Advancement of Literature and Science*. In 1886, just four years after his arrival in Carlisle, Macpherson published his first book, *The Birds of Cumberland*, jointly with William Duckworth. Although he never published under the Society's name (it was not until 1909 that the Society could afford to publish its first volume of *Transactions*), his monumental work *A Vertebrate Fauna of Lakeland* (1892) was a lasting inspiration to the young Society. This book remains a standard reference text on the natural history of Cumbria to this day. It was in this work that Macpherson outlined the boundaries of the area which he called 'Lakeland' – which closely resemble the borders of the modern-day county of Cumbria. This definition of 'Lakeland' was followed in many subsequent publications of the Society.

Macpherson was able to help the Society in another way. For several years he had enthusiastically campaigned for collections and displays of local natural history to feature in the new museum being proposed at Tullie House by Carlisle Corporation. In this he achieved great success and was appointed honorary Director of the Museum when it opened in 1893. Through personal connections Macpherson was able to secure a meeting room for the Society in the Museum. The Society first met in Tullie House on 1 March 1894 and the arrangement has continued ever since almost without break, to the mutual benefit of both organisations. Macpherson, having moved to Scotland in 1899, died suddenly at Pitlochry in 1901. His large collection of bird skins and mounts had already been given to Tullie House Museum.

One of the last projects which he undertook was the chapter on natural history in the *Victoria County History of Cumberland,* published in 1900 (Macpherson, 1900). While writing the section on vertebrates himself, he delegated the sections on the invertebrates to F.H. Day and others.

Frank Henry Day (1875-1963) was a founder member of the Society. His initial interest was chiefly in butterflies and moths, but he soon began to widen his scope to include other insect groups – particularly beetles. Along with other members of the Society, he began diligently to record the distribution of the various orders of insects found in the region. In 1900 he compiled the insect section of the *Victoria County History of Cumberland* (Day, 1900). Over the following decades he studied various insect groups and published his findings in the *Transactions* of the Society. His enthusiasm for adding new species to the county list is demonstrated by the occasion on which he found a beetle not previously recorded from Cumberland at Penton – on the Scottish side of the River Liddel. Aware that simply picking up the insect and carrying it over the border would be cheating, Day spent the next hour with a twig driving the beetle across the bridge and into Cumberland! He remained a member of the Society for 70 years, serving as Secretary for much of that time and also variously as Treasurer and President. His extensive insect collections were bequeathed to Tullie House Museum.

Initially, the Society held indoor meetings every fortnight in the winter and once a month in the summer. Members exhibited specimens and read papers which they had prepared on various aspects of natural history, some of which were later published as the Society's *Transactions*.

The Society held its first field meetings in 1894, when it visited Orton, Silloth and Rose Castle. In the first volume of *Transactions* it was explained that summer outings were arranged *'to various parts of the district, usually into some locality unfamiliar to most of the members with a view to ascertaining the birds, insects, plants etc. to be found there. In this way not a little has been added to* the knowledge of the distribution of life in Cumberland, and not a few species of insects have been added to the local lists'. It goes on to say *'These Transactions are in fact intended to be a medium of publication for all that is of interest concerning the natural history of Cumberland and Westmorland'*.

Another entomologist/collector George Bell Routledge (1864-1934) joined the Society in 1897, serving on Council in several capacities, including President on more than one occasion. Along with Day, Murray and later Harry Britten, he joined in the task of surveying the county's insect fauna. His chief interest was in Lepidoptera but, like the others, he collected and studied most insect orders. In 1897 he read an account of the butterflies of Cumberland to the Society, which he updated for inclusion in the *Transactions* of 1909. In this he acknowledged 43 species in the county excluding 'doubtful natives'. His large insect collections were given to Tullie House.

James Murray (1872-1942) joined the Society in its early years, acting as Vice-President to Macpherson in 1895/96 and as President in 1906/7. Born and educated in Carlisle, he spent a short time in Whitehaven before returning to take up an appointment with Carr's Biscuits (Day, 1942). He had a broad interest in natural history, including mosses, snails and most insect orders. Murray retired early and lived for a while at Kelsick, near Abbeytown, before moving to Gretna. Here he made many interesting discoveries – particularly of Diptera and Hymenoptera. His natural history collections were presented to Tullie House Museum by his widow.

The first volume of the *Transactions* (1909) fittingly began with a memoir of T.C. Heysham (1792-1857) by James Murray. Heysham was a gentleman-naturalist of a previous generation who might fairly be regarded as the first serious student of natural history in Carlisle. Murray made a second contribution to Volume I of the *Transactions* with a paper on the land freshwater shells of Cumberland. He wrote that Cumberland had been fairly well worked for land shells, having 63 of the 87 British species with an additional 31 freshwater species. (Today the combined total for the whole of Cumbria stands at 141 species out of around 200 British species of land and freshwater molluscs.)

Following on from the departure of Macpherson in 1899, the Society had begun to languish for want of a keen ornithologist. The entomologists continued their work but the birdwatchers began to drift away. This situation was reversed with the appointment of Linnaeus Eden Hope to the staff of Tullie House Museum in 1901. L.E. Hope was a taxidermist in Penrith during the 1880s and 90s and was commissioned by Macpherson to prepare many of the exhibits for the new displays at Tullie House. For much of this time Hope had taken a room in Carlisle from which to prepare the material and had been a member of the Society for some years. On becoming Curator, he moved to Carlisle and was able to attend meetings bringing fresh impetus to the Society. Hope did an excellent job of developing the Museum. He was also a leading member of the Cumberland Nature Reserves Association formed in 1913 to protect the wildlife of the county – Kingmoor Nature Reserve was created in Carlisle at this time and must be one of the oldest local nature reserves in the country. Hope was well ahead of his time in creating the first local biological records centre in the country in 1902. Based at the Museum, it aimed to record local wildlife sightings in order to further the work begun by Macpherson. By the time he retired in 1929, Hope was Director of Tullie House Museum and Assistant President of the Museums Association of Britain.

As a naturalist, Hope was mainly a museum man rather than a field worker (Day, 1944a). In 1907 he read a paper on the 'Gulls and Diving Birds of the Solway' in which he discussed the diet of Black-headed Gulls. He and Dixon Losh-Thorpe had examined the guts of 100 specimens following concern among fishermen that the rapidly increasing gull populations were damaging fish stocks. Their findings did not support the fishermen's claims, revealing a mixed diet of largely invertebrate food. Black-headed Gull numbers had increased enormously over the previous decade with at least 2,000 pairs at Bowness Common and 'hordes' at Ravenglass. A colony at Moorthwaite had increased from four pairs in 1879 to 1000 pairs in 1907 and this pattern was reflected at many colonies around the county. The Kittiwake, while noted as a regular visitor was then unknown as a breeding species – it now breeds in large numbers at St Bees Head. The Black Guillemot and the Fulmar were rare visitors to the Solway in 1907 – both now breed in small numbers at St Bees Head.

In 1902 F.H. Day read a paper to the Society on 'The fauna of Cumberland in relation to its Physical Geography' in which he described the diverse habitats of the county in relation to the region's geology and topography. He referred to the former forest of Inglewood which had once stretched *'from Carlisle to Westward by Thursby, thence to Caldbeck, Castle Sowerby, Mabil Cross, Blencowe and Penrith; whence its boundary ran along the River Eamont to its confluence with the Eden, which constituted its eastern limit, then northward to Carlisle'*. Also gone by 1902 were the wetland areas of Tarn Wadling to the south of Carlisle (drained in 1858), and Cardew Mire, near Dalston, where Bitterns were said to have bred and many rare insects had been found by Heysham in the early 19th century. However, many sites of interest remained, the sandstone hills running north from Penrith held many important species, while Baron Wood near Armathwaite was *'a famous locality for rare insects'*. The great wildlife value of the peat mosses was well understood by local naturalists. Day singled out Newton Reigny Moss near Penrith as of great interest, having six species of aquatic plants and 19 species of spider, together with many insects, not found elsewhere in the county at that time. Sadly, 40 years later, Day reported that much of the wildlife interest of the site had been destroyed by a large colony of Black-headed Gulls and a Starling roost (Day, 1944b). Partial drainage in the

Second World War must also have been a contributory factor. F.H. Day began his list of Cumberland beetles in the 1909 *Transactions*, dealing with the tiger beetles and ground beetles.

Through Linnaeus Hope, Day was introduced to a young gamekeeper who was keen to learn more about Lepidoptera. This was Harry Britten (1870-1954), who was the son of a gamekeeper at Skirwith Abbey and initially followed in his father's footsteps becoming head keeper at Nunwick Hall Estate, Great Salkeld. Day encouraged him to develop an interest in the study of beetles. The two soon became close friends and along with James Murray and G.B. Routledge, made many interesting discoveries regarding the invertebrates of the area. Britten joined the Society in 1902 and remained a very active member until 1913 when he left to work at the Hope Department of Entomology at Oxford University. In 1918 he moved to Manchester where, after a brief spell at the natural history suppliers Flatters & Garnett Ltd., he was appointed Assistant Keeper in Entomology at the Manchester Museum. He developed the insect collections there to be amongst the most important in the country, and became an entomologist of world renown. His own collections, including much Cumbrian material, are held by the Manchester Museum.

In 1905 Harry Britten read a paper on the 'Mammals of the Eden Valley'. He noted the absence of Polecats from the area, although some of the older residents could remember them as plentiful (a status which may soon be restored). The Weasel was becoming scarce at that time, under pressure from mole-trappers in particular. The Otter was still plentiful, while the Badger had been encountered only once in the district when one was trapped in 1888. It is worth recording that as early as 1894 at the first General Meeting of the Society, the Badger was noted as making a recovery in the area (Day, 1933), having been considered extinct a few years previously (Macpherson, 1892).

W.E.B. Dunlop of Troutbeck, near Windermere, contributed a paper on 'Westmorland Ornithological Notes, 1907'. He complained of the effect of the exceptionally inclement weather on the birds of the county, with a cold, wet and stormy spring and summer (a cry of woe familiar to all Cumbrian naturalists!). Amongst other sightings, he noted increasing numbers of Pochard on Rydal and of Pied Flycatchers near Elterwater. A single Oystercatcher at Elterwater was the first record for the district (now of course a widespread bird inland in much of the county). While reporting a sighting of six Blackcock and nine Greyhen he noted that this species was on the increase in the district. Dunlop also gives an account of an excursion to a Peregrine eyrie – the first intimation of the enduring fascination which this bird has held for successive generations of naturalists in the Society. Interestingly, after photographing the young in the nest, he noted that all the prey remains were of pigeons.

T.S. Johnstone's account of the flowers around Carlisle gave a botanical perspective to Volumes I and II of the *Transactions*. In it he noted a new record for Cumberland in Wall Whitlow-grass (*Draba muralis*) on the lane from Scotland Road to Knowefield; the Burnt Orchid (*Orchis ustulata*) was very rare, occurring at Stainton Banks and Crosby-on-Eden. Yellow Vetchling (*Lathyrus aphaca*) was found at Grinsdale in 1902 – indeed Grinsdale Gravel Beds figure prominently in the paper with many scarce and rare species. Monkhill Lough also held several interesting plants including Tubular Water-dropwort (*Oenanthe fistulosa*) and Horned Pondweed (*Zannichellia palustris*).

Tom Little Johnston (1875-1948) was introduced to the Society at an early age by his brother and founder member, Benjamin Johnston. His first paper, dealing with the wading birds of the Solway, appeared in *Transactions* Volume I. He recorded 35 species in this category, of which he had personally seen 27. In contrast to modern times, the Dotterel was fairly regularly seen on spring passage on the Solway marshes. The Lapwing was the commonest wader at that time, with 16 nests being recorded in a single field. T.L. Johnston went on to become a great field naturalist and expert on the birds of north Cumbria. He was also a keen collector, amassing an important collection of birds' eggs which he presented to Tullie House Museum in 1936 (Blezard, 1954).

In 1912 the Society was able to publish its second volume of *Transactions*. L.E. Hope contributed a memoir of H.A. Macpherson. A paper on the minerals of Cumberland read by John William Branston in 1910 was a new avenue of study for the Society. Referring to specimens in Carlisle Museum, Branston detailed the minerals occurring in the north Pennines, Caldbeck Fells, Keswick area and West Cumberland. Branston served as President of the Society in 1912/13, when the membership reached a new peak of 60.

Harry Britten contributed a paper on 'The Arachnids of Cumberland' in which he drew together his own work and that of J.C. Varty Smith of Penrith and other members of the Society with previously published lists by F.O. Pickard-Cambridge, Dr Randell Jackson and others. He recorded 301 species of spider plus 16 harvestmen and eight pseudoscorpions. The study of the county's spiders has been continued through to the present day by J.R. Parker F.Z.S., formerly of Carlisle now of Keswick, who joined the Society as a junior member in 1931 and was elected an Honorary Life-member in 1981.

E.B. Dunlop read a paper on 'The Natural History of the Peregrine Falcon' in 1912. His observation that many Peregrines remain in their breeding haunts throughout the year, and notes on the relationship between Peregrines and Ravens clearly show many hours of patient study. Dunlop kept a careful record of prey items found at Peregrine eyries – the most common at the sites he monitored being the Starling (19), followed in descending order by; Greenfinch (5); Mistle Thrush and Red Grouse (4 of each); Chaffinch, Ring Ouzel, Blackbird, Fieldfare, Redshank and Partridge (2 of each) and one each of Jay,

Carlisle Natural History Society at Castletown, Rockcliffe, 3 March 1928.
Back row (l – r): Bob Graham, Bob Brown (or James Johnstone), Ted Marriner, Mrs T.R. Stewart, Miss B. Johnson, George Walton, Ben Johnston, Ernest Blezard, Ernest Glaister.
Front row (l – r): G.B. Routledge, Miss Dobinson, Tom L. Johnston, Father Kerr, Tom Pattinson.
photo: F.H. Day junior

Members of Carlisle Natural History Society at Cumwhitton Moss, 1 May 1920.
Back (l – r): James Murray, F.H. Day, Robert Leighton, E. Day.
Front (l – r): L.E. Hope, Miss F. Hope.

Song Thrush, Hawfinch, Meadow Pipit, Cuckoo, Wood Pigeon, Woodcock, Curlew and Lapwing. In addition he refers to several instances of the remains of domestic pigeons and chickens being found.

L.E. Hope's paper on 'The Ducks and Geese of the Solway' noted the Pink-foot as the commonest grey goose on the estuary although the Bean and Greylag were almost as numerous. The latter species had increased dramatically over the previous decade so that 30% of the grey geese sold in Carlisle game shops were Greylags. Macpherson (1892) had considered it the rarest of the grey geese on the Solway. Whooper Swans were unusual visitors at this time and a single bird on the Eden at Carlisle in December 1904 was carefully observed. The Goosander was a regular winter visitor but did not breed in the county at this time, a group of seven on Talkin Tarn on 18 January 1905 was considered noteworthy (the winter of 1995/96 saw an unprecedented number of 106 Goosander counted at the same site by the present Recorder of the Society).

F.H. Day continued his list of Cumberland beetles in the *Transactions* Volume II, dealing with the water beetles and rove beetles of the county. G.B. Routledge continued his study of the local Lepidoptera with a paper on the non-geometrid 'macro' moths of Cumberland.

In November 1913 E.B. Dunlop was elected President of the Society. The following year war broke out and although the Society continued to hold meetings *'the circumstances of the time were such that interest in our affairs inevitably flagged'*. Numbers dropped off, with just two members attending the only meeting held in 1917 (Day, 1933). Dunlop was in Canada studying the incubation habits of birds at the time the war began. Although loth to leave his studies unfinished, his sense of duty brought him home in 1916. He departed for France with the Border Regiment in April 1917 and was killed in action less than a month later on 19 May 1917. This was a great loss to the Society and to ornithology in general, but most particularly to his close friend Miss Marjory Garnett of Windermere who was also a keen naturalist and member of the Society. Before leaving for France, Dunlop had left his manuscript on 'Lakeland Ornithology 1892-1913' with L.E. Hope. This paper was designed to update the information given in *A Vertebrate Fauna of Lakeland* (Macpherson, *op. cit.*) and had been read to the Society in 1913. His fine ornithological collection was bequeathed to Tullie House Museum.

With the publication of the Society's third volume of *Transactions* in 1923 came the opportunity to publish Dunlop's paper with updates by L.E. Hope. Dealing with each species in turn, Dunlop noted changes in status and other interesting records and observations of the Lakeland birdlife. He noted that the Hawfinch had greatly increased in numbers, while the Corn Bunting was at that time well-established as a breeding species in Cumberland. A Honey Buzzard was trapped in Cumberland in 1913, while a second bird seen in the same locality for some time after suggested that a breeding attempt had been foiled. The Greylag Goose was by this time the commonest grey goose on the Solway, *'equalling all others in numbers'*. The Brent Goose, although far from plentiful, occurred on the Solway in most winters at this time. Barnacle Geese were thought to be on the increase despite attempts to reduce their numbers. The Corncrake, already on the decline in some parts of Lakeland, still abounded on the west coast near Drigg. The Green Woodpecker was a recent arrival as a breeding bird in central Lakeland, while Turtle Doves nested frequently near Carlisle during this period. The Common Gull was reported nesting in England for the first time when a pair bred at Long Newton Marsh in 1914. The second Sabine's Gull recorded for Cumberland was *'obtained'* at Anthorn in October 1921 and provides an early reference to oil pollution. The bird was found to be *'covered above and below in sticky oil and tar, as were many other birds at this time'*.

Transactions Volume III also included the third part of Routledge's paper on the Lepidoptera of Cumberland, dealing with the Geometrae. He listed 203 species for the county and acknowledged the help of Day, Britten and Reverend Harold Dodsworth Ford in putting together the list. Day's list of the Coleoptera of Cumberland concludes in this volume, with the total number of beetle species in the county given as 1,797 (this was later updated to 1,837 in Volume V).

Ernest Blezard (1902-1970) joined Carlisle Natural History Society as a young man. He made his first contribution to *Transactions* with a paper on the Raven, read in 1927. He gave an interesting account of the life history of the bird, including its choice of breeding sites, persecution by gamekeepers, interaction with Peregrines, and colonial roosting. Blezard referred to trips made with Ritson Graham, who also contributed to this volume of the *Transactions*, with a paper entitled 'Local Wildfowl', first read in 1926. Graham noted that the Bean Goose had by then become very scarce on the Solway. The increase in Pochard numbers first noted in 1907 (Dunlop, 1909) was continuing – Graham estimated a doubling of numbers over the previous six years. He speculated that this bird would soon be added to the list of breeding species for the area and proved himself right by finding two nests with eggs in the spring of 1927. The Tufted Duck was a recent addition to the list of breeding species with three sites known and some evidence of others. Smew were thought to be increasingly frequent winter visitors at this time, having previously been considered an *'irregular and uncertain visitor'* (Macpherson, *op. cit.*) – Monkhill Lough held several birds in 1924/25. Sadly this water is now drained.

Routledge concluded his list of the Lepidoptera of Cumberland in *Transactions* Volume IV, covering the families of 'Micro-lepidoptera'. In the same volume he contributed a paper on the Orthoptera of Cumberland, listing 14 species for the county of which three cockroaches and the House Cricket were non-native synanthropic species. (Two of the remaining 10 species were probably recorded in error). The Bog Bush-cricket (*Metrioptera brachyptera*) was common on

Wan Fell in 1900 but appears to be extinct there today (It has since been recorded from Wedholme Flow in 1991 and is known from several mosses in the south of Cumbria). Routledge suggested that Speckled Bush-cricket (*Leptophyes punctatissima*) which had been reported from Wigtownshire in 1906 might occur in Cumberland. In this he was eventually proved correct, when in 1994 another member of the Society, Roy Atkins, discovered the species at St Bees Head in company with Dark Bush-crickets (*Pholidoptera griseoaptera*) – also additional to Routledge's list.

F.H. Day published his paper on the Hemiptera-Heteroptera (land and water bugs) in this volume of the *Transactions*. He listed 214 species present in Cumberland out of 475 British species – largely the result of his own work together with Britten, Murray and Routledge. Today, including the rest of Cumbria, the total stands at some 280 species out of around 560 occurring in Britain. Most of the species listed by Day are still known in the county. Some appear to have decreased in abundance – Day regarded *Strongylocoris leucocephalus* as '*rather common in the sweep-net*' in contrast to the species' present local distribution. This particular decline is probably due to the loss of unimproved grassland habitat. In addition several species recently added to the county list may well have extended their ranges from the south since Day's publication. The sand dune dwelling stilt bug *Chorosoma schillingi* was first found in Cumbria at Sandscale Haws by Dr Neville Birkett in 1975. It is now additionally known from Eskmeals and Drigg dunes. The latter site was a favourite collecting ground of Day's and he is unlikely to have over-looked such a large and distinctive insect. Other examples include the Hawthorn Shieldbug (*Acanthosoma haemorrhoidale*) which was not included in his published list, however his own annotated copy (now held by W. Kydd of Ulverston) records the species first taken at Ravenglass and Pooley Bridge by W.F. Davidson in the late 1950s. This species is now widespread in the county and has continued its spread north to the Scottish Highlands.

In 1933 the fifth volume of the *Transactions* was published. F.H. Day gave a first-hand account of the first 40 years of the Society in which he noted that attendance at that time was very good, with numerous young members of promise. Two papers by Benjamin Johnston (who joined the Society in 1894 and twice served as its President) record first the birds, then the mammals found within the city of Carlisle. Of the birds he had recorded 112 species, of which 52 had been noted nesting. Although Corn Buntings were not common in the city several pairs were known to nest nearby. Among the species which had nested in the city, the Whinchat was noted as formerly plentiful in the Denton Holme district where it had nested freely at the top of Denton Street, while the Barn Owl was still '*fairly well distributed*' and the Corncrake, though less common than it had been, had reared young in the previous two seasons (1929 & 1930) not far from Crown Street. Of the mammals, 17 species had been recorded within the city. Otters occurred on the Eden and Caldew and were known to breed within the city – a situation to which we have happily returned in recent years. Water Voles while frequent outside the city were not common in Carlisle where it was considered they might be displaced by Common Rats.

The Reverend Harold Dodsworth Ford, Rector of Thursby, became President in 1929. A keen lepidopterist, he contributed a paper on 'Collecting at Light in the Carlisle District'. Having used an acetylene lamp to attract moths at the Rectory over the previous 20 years, Ford was able to give useful advice on the most effective technique and also included a list of the 204 species which he had recorded by this method. Ford's son was also a member of the Society in later years. He was E.B. Ford, the famous geneticist and author of the *Moths* and *Butterflies* volumes in the Collins *New Naturalist* series. Their combined collections (containing important Cumbrian material) are now in the Hope Department of Entomology at Oxford University.

A paper on the 'Trees in Carlisle' was the first by the botanist Dorothy Stewart, later to become Mrs Ernest Blezard. Having joined the Society in 1923 Dorothy, now an Honorary Life-member, is the longest serving living member of the Society – with memories going back to L.E. Hope and other early luminaries of the Society. In the same issue Ernest Blezard contributed a paper 'On the Buzzard', in which he presented information gathered in the field with Ernest Glaister and Ritson Graham.

The last published contributions from G.B. Routledge before his death in 1934 took the form of an Appendix to his previously published lists of Lepidoptera in which he adds 34 species to the county list. In this the name of Canon G.A.K. Hervey appears against a number of records from Buttermere where he had lived from 1926 to 1931. (Hervey was later to found the Lake District Naturalists' Trust.) His collection of Lepidoptera was given to the present Cumbria Wildlife Trust in 1969 and is now held at Tullie House Museum. A paper on the Neuroptera and Trichoptera of Cumbria and another on the Aculeate Hymenoptera of Cumberland also appeared under Routledge's name in Volume V.

In a paper on the Barnacle Goose, T.L. Johnston examined the changes on the saltmarsh of the English Solway likely to have caused the observed decline in Barnacle Goose numbers at that time. He noted that the geese preferred to feed on the fine young marsh grass and glasswort found on accreting saltmarsh. Looking at each marsh in turn he recorded the historic use of the marsh by the geese and the changes between accretion and erosion which had taken place at many sites, causing the birds to move to new pastures. While Rockcliffe Marsh had once been a favoured site for Barnacle Geese, their numbers began to reduce in the 1880s as better feeding grounds were created on the accreting saltmarsh at Long Newton. Early in the 20th century, changes in the course of the Eden caused the lower marsh at Rockcliffe, where the geese had traditionally grazed, to be washed away. Long Newton Marsh continued to be one of the best sites for Barnacle Geese for many years – here on 12 December 1892 the punt-gunner William Nichol of Skinburness killed a record 40 birds with one shot! Skinburness Marsh, which had also been a good place for Barnacle Geese had latterly stopped growing and the vegetation had become rank and unpalatable to the geese which had practically deserted as a result. The paper also includes an account of one of the early expeditions to the Barnacles' Spitzbergen breeding sites in the 1920s.

'The Roe Deer in Cumberland' was the subject of a paper by Ritson Graham who reviewed its history and detailed its expanding distribution and numbers, which Graham considered to date from the period of The Great War.

It would be difficult to over-state the contribution made by Ernest Blezard to the study of natural history in Cumbria through his work for the Society and Carlisle Museum where he worked for 40 years. First serving as Librarian to the Society in 1926, Blezard went on to serve in every capacity over the next half century, including President from 1968 to his death in 1970. In 1939 the manuscript to *The Birds of Lakeland* was ready. Intended as a thorough review of the status of Cumbrian birds, it could not be published at that time. Despite the war-time difficulties this, the sixth volume of *Transactions*, appeared in 1943 – the Society's 50th anniversary year! Edited by Blezard and co-written with Marjory Garnett, Ritson Graham and Tom L. Johnston, this acclaimed publication considered some 290 birds for Cumbria. Chief among the changes since Macpherson's publication in 1892 were the additional sea-bird species breeding at St Bees Head. The earliest record of Kittiwakes breeding was at St Bees in 1932 when some 20 pairs were counted – numbers were reported to be gradually increasing. Fulmars were an exciting addition to the county's breeding birds when Ralph Stokoe concluded that one of the five pairs present at St Bees in 1940 had laid an egg. Black Guillemots were also first reported breeding at St Bees in 1940, when three pairs were noted by Ralph Stokoe and Austin Barton.

Lakeland Natural History was the title given to Volume VII of the *Transactions*, published in 1948. In this volume Flt. Lt. R.A. Carr-Lewty gave an account of the observations of bird flight paths and behaviour in the county, made possible from an aircraft. A study of the foraging behaviour of Rooks at Kingstown Carlisle was made in this way. Foraging was recorded up to three and a half miles from the rookery, often overlapping ground covered by adjacent rookeries. Interestingly it was noted that the rooks tended to fly out into the wind to feed, returning down-wind to the rookery. On calm days the birds foraged randomly away from the colony. The flight movements of birds in relation to topography and tides on the Solway were also discussed and there was reference to the altitude and flight speed at which various flocks of birds had been accurately recorded from aircraft.

Marjory Garnett (1896-1977) was born in Windermere and lived variously at Kentmere, Cartmel, Seascale and Kirksanton where she died (Blezard, 1983). She was introduced to the Society through her friend Eric Dunlop, remaining a member throughout her life. A keen conchologist with particular interest in the marine species of the county, her first love was ornithology. With the latter study she combined her artistic talents, illustrating the soft parts of specimens in her extensive collection of study skins. This collection and the associated watercolour illustrations was given to Tullie House Museum in 1936.

In *Transactions* Volume VII Miss Garnett contributed a retrospective study of winter birds on Lake Windermere – 1912-33. She recorded that Cormorants first roosted on the lake in the winter of 1917/18 and noted that Coots had formerly nested in the reedy margins, but that they had ceased breeding as the reedbeds became reduced in size. Tufted Duck and Goldeneye both increased in numbers over this period, while the felling of many oak trees on the islands in World War I reduced the numbers of woodland birds. Great Crested Grebes began nesting on Esthwaite Water in 1908 and occasionally on Blelham and other tarns. The hard weather in 1929 brought many interesting birds to the lake including Blackthroated Diver, Slavonian Grebe, Red-breasted Merganser, Smew and Scaup. Black-throated were the most frequent of the divers on the lake between 1915-1930 with Red-throated being the scarcest.

In an account of 'Creeping Lady's Tresses in Cumberland', Dorothy Blezard mourned the loss of all but one of the known colonies due to woodland clearance. Concern over the possible local extinction of the species was tempered by the hope that it would colonise and survive in some of the many new plantations then being created. (The species persists to the present day in a few plantations and semi-ancient sites.)

A review of the 'Greylag Goose in Lakeland' was contributed by T.L. Johnston. He commented that historical records of geese breeding at Sunbiggin Tarn were attributable to the Greylag. Previous authors in the 19th century had consistently regarded the Greylag as the rarest grey goose in the region. Johnston dealt with changes in the Solway feeding grounds over the first half of the 20th century. He noted that the increase in Greylag numbers matched the increase in available grazing marsh for the geese. Having first increased in numbers on the Scottish shores of the Solway, the Greylag was the commonest goose in Cumbria by 1918, but was soon deposed by the rapidly increasing numbers of Pink-footed Geese. From 1935 Greylag numbers had declined on the Solway, transferring to the south shore of Morecambe Bay.

The history and status of the Badger in Cumberland was detailed by Ritson Graham. He argued that although seriously reduced in numbers towards the end of the 19th century the Badger had never been entirely exterminated from the county. A steady recovery was noted from the turn of the century with a rapid increase about the time of World War 1. He noted just two occurrences of the rare erythristic colour phase in Cumbrian Badgers – one at Bewcastle and the other from the Castle Sowerby district. At the time of writing the Badger was just returning to the area of the central Lakeland fells. Graham noted that Badgers were falling victim to the increasing road traffic, with instances invariably being reported in the local press – such road casualties are now so common as to excite little comment.

In a supplement to *The Birds of Lakeland,* Blezard included a record of Golden Oriole in Baron Wood in 1943 and the first account of Siskin breeding in Westmorland. The occurrence of the Nuthatch at Seathwaite, Dunnerdale was considered noteworthy, while the Green Woodpecker was noted to be expanding its range into east Cumbria.

Tom Little Johnston died in 1948 – an obituary appeared in Volume VIII of the *Transactions,* published in 1952. His last paper, on the Great Black-backed Gull on the Cumberland Solway, was published posthumously in the same volume. In this paper he detailed the history of the bird, which bred in low numbers on several of the south Solway mosses. Extensive fires on Bowness Moss in 1927 and again in 1942 were referred to – the former being regarded as the reason for the re-location of the Lesser Black-backed Gulls from Bowness Common to Rockcliffe Marsh in 1928. After the large-scale colonisation of Rockcliffe Marsh by Lesser Black-backeds, it became usual to find one or two pairs of Great Black-backed Gull there too – six pairs in 1934 being the highest number recorded at that time.

A paper by Mathew Philipson entitled 'North-eastern Bird Studies' noted that while the Black Grouse had previously been common in north-east Cumberland, it had declined dramatically and was by then very scarce, with fears for its local extinction. Robert H. Brown contributed a paper on the breeding warblers of the county in which he regarded the Whitethroat as almost as common as the Willow Warbler at that time – sadly not the case today.

The newly notified Site of Special Scientific Interest at Siddick Pond, Workington was the subject of a paper by Ralph Stokoe. Its history and environment were described and an annotated list of species included reference to the first Cumbrian record of the Little Bunting, seen there in 1948.

1958 saw the publication of Volume IX of the *Transactions,* again edited by Blezard, containing a third supplement to *The Birds of Lakeland,* and an extension to the article on the food of birds published in the previous volume. Blezard noted the decline of the Grey Wagtail at the time (*'apparently following in the way of the Stonechat'*) as well as the Long-eared Owl. Conversely, he noted an increase in Song Thrush and Blackbird numbers, the Green Woodpecker and Curlew were spreading, and he stated *'the Corncrake has by no means yet vanished'.* Carrion Crows, gulls and oceanic birds on the sea cliffs were also increasing. Referring to the crash in rabbit numbers due to Myxomatosis, Blezard contended that Buzzards were increasing in range and numbers within the county, having switched to preying more on other birds and carrion, and also, in 1957, to the then abundant Field Vole. This minor 'vole plague' also benefited Short-eared Owls and Kestrels at that time.

Ralph Stokoe (1921-1981) became a major figure in the Society and in the study of various branches of natural history in Cumbria. He was a founder member of the Lakeland Naturalists' Trust (now Cumbria Wildlife Trust), ultimately becoming its Vice-chairman. He studied botany, being particularly interested in orchids which he photographed in Britain and Europe. Later he studied ferns and then aquatic plants – systematically surveying every lake and tarn in Cumbria (Halliday, 1982). Stokoe became involved in the Flora of Cumbria Project organised by Dr G. Halliday, contributing records from his home area of Cockermouth. His first and greatest love however was birds. He served on Council of the Society for many years from 1950, and in 1962 edited the tenth volume of the *Transactions,* entitled *Birds of the Lake Counties.* Several 'new' breeding birds for the county appeared for the first time in this publication. The Golden Eagle had returned after nearly two centuries' absence; the Collared Dove had arrived and the Eider, Goosander, Red-breasted Merganser and Crossbill were also recent colonists.

Raymond Laidler (1918-1994) was a contemporary and great friend of Stokoe, sharing his fascination for orchids and birds. Laidler joined the Society in 1933, learning his natural history among the woods and mosses of the Solway Plain. However, it was the hills of Galloway that he grew to love in later years. He was one of the first to notice the decline in raptor populations in the 1950s and 60s due to pesticide poisoning, and did invaluable work monitoring the populations of Peregrines and Sparrowhawks in Cumbria and south-west Scotland for the Nature Conservancy (Horne, 1994). Laidler also studied butterflies, moths and dragonflies, contributing many records to the new national recording schemes in the 1960s and 70s. He first served on the council of the Society in 1961, assuming the duties of Secretary the following year. Over the three decades he served the Society variously as Secretary, Treasurer and Vice-President, being elected President for three years from 1970.

Volume XI of the *Transactions*, entitled *Lakeland Molluscs* was published in 1967. Marjory Garnett and Mary M. Milne contributed 'A preliminary list of the marine mollusca of Lakeland', while Ernest Blezard wrote on the distribution of the non-marine mollusca, and on molluscs as the food of birds.

Robert Henderson Brown (1901-1980) of Cumdivock near Dalston was another fine ornithologist and member of the Society at this time. Educated at Carlisle Grammar School he began studying birds in his teens. Brown first served on Council in 1949 (elected President in 1964-68) and in the January of that year

Ralph Stokoe (with Major J. Rose) counting nests at Ravenglass gullery, ca 1969. *from a photograph by Brian Spencer*

Raymond Laidler in the Galloway Hills, April 1978.
from a photograph by Geoffrey Horne

had addressed the Society on 'Some results of ringing ten thousand birds'. He was one of the first to join the Witherby Bird Ringing Scheme, later taken over by the B.T.O. Robby Brown is frequently cited as the authority for various information (especially fledging periods) in *The Handbook of British Birds* (Witherby *et al.*, 1938-41). He long held the record for the longest serving ringer in the country (Horne, pers. comm.), ringing many thousands of birds from Golden Eagles to Wrens. As a Milk Recorder for the Milk Marketing Board he travelled the local farms and had the opportunity to ring several thousand swallow nestlings (895 broods – averaging 4.18 young per brood). Brown's *Lakeland Birdlife 1920-1970* has many parallels with the Society's *Transactions*. It is very much his personal account of the changes in the avian fauna of the county over a 50 year period, drawing on his own observations and ringing activities. Sadly, as he destroyed his records and note books before his death, *Lakeland Birdlife*, along with his ringing returns held by the B.T.O., is all that remains of his life's work. His natural history library was given to the Society.

Ritson Graham (1896-1983) joined the Society in the 1920s, contributing several papers to the *Transactions* over the years, and after first serving on the Society's Council in 1928 went on to serve in various capacities including President from 1931-33. Graham was a locomotive driver on the railways, but always said he would have liked to have been a gamekeeper. He was a councillor of Carlisle City for 25 years, serving as Mayor in 1956-7. He gained most satisfaction from his duties on the Public Library and Museum Committee: Tullie House with its wonderful natural history collections, cared for by his friend Ernest Blezard, was part of his life. He spent most of his spare time out in the countryside, on the marshes or among the Bewcastle Fells. It was fitting that in the centenary year of the Society for which he did so much, his manuscript notes on the Bewcastle area were finally published in book form (Graham, 1993).

Today the Society continues to pursue its studies of the natural history of the region. An active membership includes some long-time members with memories of the Society as it was just after the war and, in some cases, right back to the 1920s and 30s. Younger members learn from and build on the work of their predecessors in various fields of study including botany, fungi, birds, mammals, mollusca, reptiles and amphibians as well as many different insect groups.

From about the mid-1970s membership has increased to around 100, with attendances at indoor meetings varying between about 45 and 90. The relatively high turnover in members first noted by F.H. Day (1933) still continues with several new members joining each year, many of whom subsequently drift away. Indoor lecture meetings have maintained a varied approach covering a wide diversity of subjects. The programme also includes, very successfully, two Members' Nights per season, at which members give short accounts of their current interests and studies. Since 1992, practical workshops on specific themes have been developed and have attracted new, as well as existing, members.

CNHS Fungus Foray at Talkin Tarn, 28 September 1996, led by Geoff Naylor *(centre right).*

Topics have included sedges, ferns, dragonflies, hoverflies, freshwater invertebrates, snails and shorebirds.

The occasion of the Society's centenary in 1993 was celebrated with a very successful conference on the changes in the county's natural history over the preceding 100 years. After almost three decades in which, while contributing actively to various county publications, the Society did not publish under its own name, the centenary has seen a renewal of publishing activities, with the initiation of the *Carlisle Naturalist* in 1993, and with the publication of this twelfth volume of *Transactions*.

The future of the Society looks secure, and it is to be hoped that it will continue to prosper and actively contribute to the knowledge and appreciation of the wildlife of a very special area.

References

Blezard, D., (1983), Unpublished biographical note on Marjory Garnett. Tullie House Museum file.

Blezard, E., (1954), Tom Little Johnston 1875-1948. *Transactions of the Carlisle Natural History Society* **VIII**: 5-7.

Day, F.H., (1900), Insects. Pp. 101-142 in *Victoria County History of Cumberland* Vol 1. Repr. 1968, London: Dawsons.

Day, F.H., (1933), History of Carlisle Natural History Society. *Transactions of the Carlisle Natural History Society* V: 1-13.

Day, F.H., (1942), Obituary: James Murray F.R.E.S. *The North Western Naturalist* XVII: 115-6.

Day, F.H., (1944a), Obituary: Linnaeus Eden Hope. *The North Western Naturalist* XIX: 180-2.

Day, F.H., (1944b), Notes on some Cumberland beetles. *The North Western Naturalist* XIX: 128-32.

Graham, R., (Eds: Matthews, S. & Clarke, D.J.), (1993), *A Border Naturalist: the birds and wildlife of the Bewcastle Fells and Gilsland Moors 1930-1966.* Carlisle: Bookcase.

Hope, L.E., (1909), Gulls and Diving Birds of the Solway. *Transactions of the Carlisle Natural History Society* I: 75-97.

Hope, L.E., (1912), H.A. Macpherson, MBOU – a memoir. *Transactions of the Carlisle Natural History Society* II: 1-13.

Horne, G., (1994), William Raymond Laidler 1918-1994. *Carlisle Naturalist* 2 (1): 16.

Macpherson, H.A., (1892), *A Vertebrate Fauna of Lakeland.* Edinburgh: David Douglas.

Ernest Blezard (1902–1970)
from a photograph by W.F. Davidson, 1943

ERNEST BLEZARD 1902 – 1970

Derek Ratcliffe

Ernest Blezard was born in 1902 at Greenodd in Lancashire North of the Sands, though when he was seven his parents moved to Carlisle, where he spent the rest of his life, and which became, by adoption, his native town. There, his youthful enthusiasm for the wild creatures of the countryside developed, and was fostered by rewarding friendships with local naturalists of the old school, notably veteran wildfowler-fisherman Thomas Peal, celebrated sportsman Robert Raine, and ornithologist Tom Johnston. There were few openings then for a young man aspiring to a career in natural history, particularly when his parents had little money or social influence. He began working as a draughtsman in the government munitions factory at Gretna, but recognition of the young naturalist's abilities was such that in 1926 he was invited to become assistant to the then Director of the Carlisle Museum at Tullie House, Linnaeus Eden Hope, a post which he gladly accepted. He was taught taxidermy by Bob Raine and John McHardy – a skill which he continued to develop. On Linnaeus Hope's retirement in 1929, Ernest took charge of the natural history work, being later styled as Keeper of Natural History and Post-Mediaeval Antiquities, and remained at Tullie House until his retirement in 1966.

For the naturalist, Carlisle is a marvellous centre, lying within reach not only of the Lakeland fells, but also the moorlands of the Pennines and the Borders, and the Scottish hills of Langholm, Moffat, Nithsdale and Galloway. Then there are the great grass salt marshes and sand flats of the Solway, sand dunes, shingle beaches and the sea-cliffs of St Bees Head; peat mosses of the lowlands; lakes and tarns; river valleys and woods of widely varying character. Ernest Blezard spent a lifetime exploring all of these, in responding to the challenge of their endlessly fascinating wildlife and the thrill of their natural beauty. In so doing he acquired a now rare breadth as well as depth of knowledge in the realm of natural history.

He became a specialist in ornithology, well grounded in the formalities of bird biology and taxonomy, but with a unique combination of field and museum experience that gave him an extraordinarily profound knowledge of birds and their ways. His skill at identifying single feathers, presented to him out of the blue, was a constant source of amazement to others. He also knew a great deal about other vertebrates, the main groups of invertebrates, and the flowering plants. While his field experience was gained mainly in his beloved Cumberland and its adjoining counties, he had read widely as regards national and international background. He became a Member of the British Ornithologists' Union, a Fellow of the Zoological Society, and a Founder Member of the British Trust for Ornithology.

Ernest Blezard's influence as professional and amateur is difficult to separate. In Tullie House, his work was on show to reveal taxidermy as a fine art and, although – regrettably – the remarkable Bird Room in Tullie House no longer exists as such, some of his best examples of nesting groups remain on display to make the point. The other splendid reference collections of biological material were arranged and catalogued by him with meticulous care. His local field knowledge allowed him to portray the Lakeland natural history scene in the public displays, and he helped to give Carlisle the distinction of possessing one of the finest provincial museums in Britain. Dissection of the many birds brought to the Museum led to an interest in the food they contained, and over the years Ernest became an expert on the food of birds. His published and unpublished records on the subject are a tribute to the skill with which he applied his knowledge of the plant and animal kingdoms in identifying gut contents and other remains, and he was generous indeed in helping other ornithologists with these time-consuming analyses, often continued late into the night. He seldom collected material for the Museum himself, since it was well supplied otherwise, but now and then took himself off into the field officially to model some habitat intended to figure in a display case.

Another major interest was in 'Byegones' (his expression), and he was well versed in local history, besides having to cope, at times, with all manner of man-made collectors' items, including clocks, furniture, porcelain, books, weapons, coins and stamps. He was his own boss in programming work, and successive visits to his inner sanctum could find him setting up a fine specimen of a Greenland White-fronted Goose from a local wildfowler on one occasion, and arranging an exhibit of leather bottles the next. Now and then, he appeared as an expert witness at treasure trove inquests, though when people turned up wanting to know how much their prized possessions were worth, they were politely referred to commercial valuers.

Ernest's museum influence lay equally in personal contact with visitors. A strong sense of duty to his public, and a lively interest in his callers, made his room an information 'clearing house' for north-west England, and gained him a wide circle of friends. He was teacher, sage and helper to countless seekers after knowledge who reached his den, from small boys clutching boxes with cherished birds' eggs to university scientists bent on their local researches. And from naturalist friends who dropped in for a 'crack', there came many a choice item of wildlife news. So, he learned much from his visitors, too, whilst being mentor to so many of them. The displays in the public galleries were, nevertheless, his most visible achievement. Museums fulfil a vital educational role, and no one did more than Ernest to make Tullie House a major cultural institution of north-west England.

Yet the field meant more than the work bench or office desk to Ernest Blezard. The bicycle was first and last his means of transport, carrying him in his youth on before dawn starts for the Solway and its wildfowl, or to the fell country where he would often sleep rough, in disused mine buildings, shepherds' huts or even an open sheep-fold. The spectacular crag birds, Peregrine, Raven and Buzzard had always a magnetic appeal, but nothing gave him greater pleasure than encounters with the charming and elusive Dotterel of the windswept high tops. It was his ambition, and one almost realised, to find the nest of every bird breeding in the Lake Counties. He grew up at a time when egg collecting was still the norm, and remained until the end an unrepentant collector, though his collection was modest and carefully chosen, and he deplored the 'big series' mentality. Many a time he was satisfied simply to have found the nest and recorded its contents, coming home only with his memories, which he said made the best kind of collection of all. The outwitting of difficult subjects gave him the greatest satisfaction, as in the tracking down of elusive nests of species such as Water Rail, Crossbill, Wigeon and Short-eared Owl. His natural tree-climbing ability was an asset, but patience and persistence, a good ear and a keen eye, were all brought to bear in the quest.

Ernest was fortunate to meet Dorothy Stewart, an enthusiastic botanist, whom he married in 1933 and who could share the pleasures of the field, looking at the lovely plants of the Cumbrian countryside, as well as the animals. When Dorothy could not accompany him, he would take choice plants home for her to see and grow and paint. She delighted in the garden at their Blackwell home, and visitors had many an enjoyable and instructive tour as their hostess enlarged on the treasures she so carefully tended. Dorothy was always welcoming to Ernest's friends and interested to hear their news, especially on matters botanical. Together they built up a great knowledge of the local flora, and for many years she ran a summer exhibit of the more common wild flowers at Tullie House. Dorothy presented to the Museum in Ernest's memory a set of the *Drawings of British Plants* by Stella Ross-Craig, many of which she had accurately hand-coloured herself from living material.

A field outing with Ernest was a delight, not only for the gleanings from his lore of fieldcraft and wildlife, but also from his fund of whimsical reflections and choice anecdotes. These were told in inimitable style, with a neat and original turn of phrase and many a humorous touch and twinkling eye, between puffs of his inseparable pipe. He was no lover of large field meetings, greatly preferring to be with one or two boon companions, and quite often worked alone. Latterly he was glad to join motorised friends in trips to more distant places, and when his three sons Peter, Andrew and Crispin followed the modern trend, family outings by car became a great feature of his later years.

Ernest Blezard was for many years Secretary of the Carlisle Natural History Society and diligently worked to maintain the traditions of scientific enquiry established by its founder, the Reverend H.A. Macpherson. It was an institution he revered and, with Dorothy's help, he became its driving force. I well remember how, as a shy small boy, I was welcomed and encouraged at the meetings. They were held in the Jackson Library where the Herbarium was also housed – a place of quiet scholarship which is also, sadly, no more. It was wartime and sometimes only a handful of members attended, but the meetings were full of interest, and there was always an occasion when noteworthy specimens were passed around for scrutiny. The highlight was when the great wood and brass lantern projector was wheeled out by the Secretary for a showing of large glass slides.

Ernest was elected President of the Society in 1968 and held this office at his death in 1970. He also acted as Editor of the *Transactions*, a role which he developed energetically and which saw publication in 1943 of *The Birds of Lakeland*, by a team of authors under his leadership. A model regional avifauna, produced under difficult wartime conditions, this was probably his proudest achievement. It was singled out for praise by James Fisher in an essay, *The Last Hundred Bird Books*, in 1948. Ernest used to say that, had he known beforehand of its success with the public, he would have aimed at a larger and more ambitious treatment, which the material warranted. Various supplements were issued in later volumes of *Transactions* of the Society that the Editor saw through the press.

Ernest did not write a great deal, but all his papers have the mark of quality, from early contributions on Raven, Buzzard and Dotterel, to later studies of the food of birds and other topics. His last work was a small volume on Lakeland molluscs, conchology being another of his interests. One disappointment was the abortive project for a *New Naturalist* volume on Lakeland, another team venture: the Collins Editors praised his own chapters on birds and mammals, but judged his colleagues' efforts short of the mark. The alternative suggestion, of co-authorship with an academic scientist, did not appeal, and the enterprise fell through. He took great pains in writing, and developed a distinctive style: some of his shorter essays in the local press and magazines were little gems, sadly forgotten but deserving resurrection.

Throughout his life, he kept a detailed account of all his more significant field trips, and these journals, now housed with the County Archives, remain as a most valuable record of wildlife in Lakeland and elsewhere. They are, beyond this, a moving and inspiring statement of what can be achieved by determination when few opportunities or facilities are on hand, and two weeks annual leave was the longest free-time spell available. What could he have accomplished with some of the opportunities for indulging fieldwork that we nowadays take for granted?

Ernest followed the post-war development of the nature conservation movement with great interest. When, in the early fifties, the recently created Nature Conservancy began its programme of notifying Sites of Special Scientific

Interest, he was asked for his views on suitable localities and for help in surveying these. He journeyed out on his bike to places such as Rockcliffe Marsh, Biglands Bog, Cumwhitton and Moorthwaite Mosses and Bolton Fell, and provided neat accounts of their habitat and wildlife interest. Having diagnosed the cause of occasional bird 'kills' by agricultural chemicals long before the organo-chlorine insecticide episode, he was much concerned over such problems and, for instance, helped Ian Prestt with his surveys of Sparrowhawks in relation to these pesticide effects in the Solway region. His observant eye also took in some of the other impacts of modern farming practice – the decline of Lapwings and Corn Buntings near home, the ploughing and 'improvement' of Snipe and Redshank meadows in the marginal lands, and the disappearance of once colourful floral displays of orchids and Bird's-eye Primrose on grasslands treated with basic slag.

In matters of observation and data, his standards were impeccable, and any suspect records were ruthlessly eliminated. He always held that this is something where no mistakes or deviations are allowable – the price is a tarnished reputation forever. Ernest's principles were high in every way, and he had little time for those who fell appreciably below them. In his later years, he also felt increasingly on a different wavelength from the modern world of ornithology, with its graphs and statistical analysis, impenetrable prose and scientific remoteness. This led him to withdraw and go his own way. He identified with an older generation of naturalists, and with the ordinary, simple countryman; and the modern world in general embodied much that was distasteful to him. Yet in many ways he was not resistant to change and, for instance, had warmer feelings towards the Forestry Commission and its works than do many naturalists.

This remarkable naturalist, whose encyclopaedic knowledge was largely self-taught, hid much of his talent under a bushel when it came to conventional advancement. He declined an appointment to the prestigious York Museum, having found that normal working conditions excluded natural daylight, and having gazed upon the unappealing flatness of the Vale of York. Ernest craved the simple life and had no aspirations to fame or wealth, finding satisfaction in the job well done, and in good human relationships. He remained an eager seeker after knowledge in his own way, but perhaps even more his passion was in, to use his own phrase, the 'simple unalloyed pleasure' of seeing Wild Nature for its endlessly magical beauty. His philosophical bent expressed itself once in mild rebuke. We had scrambled for hours up the great scarp of Wasdale Screes, enjoying the sight of choice plants on the way, and finally arrived on the top. I began looking at my watch and wondering if the day might still be young enough to fit in some other interesting spot. Ernest had made himself comfortable against a rock and was packing his beloved pipe; said he, "Sit down, man, and absorb something of the peace of your surroundings".

It is fitting that his last field trip, only four days before his death, from a heart attack on 4 April 1970, was with a kindred spirit, Geoff Horne, to see the Golden Eagles which had recently returned to breed in their ancient haunts among his native Lakeland fells. His loss was so premature and irreplaceable, but it is good that he lived to witness this crowning event in Lakeland ornithology. In concluding this brief memoir, I acknowledge my own debt to Ernest Blezard, as my mentor and friend, and shall always regard him as an unsung genius.

Contributors

David J Clarke

Senior Curator & Collections Manager at Tullie House Museum Carlisle, and President of Carlisle Natural History Society. A natural historian by training, he is currently North of England Recorder for the national Odonata Recording Scheme. He has lived in Cumbria since 1969 and has a wide knowledge of its dragonflies and their habitats.

Geoffrey Halliday

Noted botanist and Senior Lecturer at Lancaster University's Institute of Environmental & Biological Sciences. He has been the driving force behind the 'Flora of Cumbria' project. As a Council member of the Cumbria Wildlife Trust, and Chairman of its former Scientific & Advisory Committee, he has for many years played a key role in local conservation matters.

Stephen M Hewitt

Keeper of Natural Sciences at Tullie House Museum Carlisle, Secretary of the Carlisle Natural History and organiser of its centenary Conference. A local-born zoologist and keen general naturalist, he has been instrumental in the recent development of the Tullie House Local Biological Records Centre.

Geoffrey Horne

Carlisle naturalist and ex-President of its Natural History Society. He has carried out extensive studies of birds of prey in Lakeland over the past 30 years, notably on Peregrine Falcons. He has recently been awarded the prestigious RSPB Medal, and the Bernard Tucker Medal, for his lifetime contribution to ornithology.

Roger Key

Invertebrate Ecologist with English Nature, he has frequently appeared as a radio broadcaster on his wildlife specialities. His expertise on beetles and other 'lower animals' has been ideally suited for compilation of the national Invertebrate Site Register – in which Cumbria features extensively.

Geoff Naylor

Local naturalist and Recorder for the Carlisle Natural History Society. He has been recording wildlife for as long as he can remember and has particular interests in butterflies and moths. In various capacities, he has been responsible for data-entry at the Tullie House Biological Records Centre.

Derek A Ratcliffe

Has led a distinguished career in nature conservation, retiring as the Nature Conservancy Council's Chief Scientist in 1989. He is also widely known as the author of many seminal books and papers on the British flora and fauna. His unrivalled knowledge as a field naturalist, and special associations with Carlisle and Cumbria, have ideally suited him to review developments in the county.

John Webster

Has had a long-standing involvement in the Mammal Society (of which he is a Council member) and has specialised in the mammmals of his native county and their conservation. His in-depth studies of Cumbrian mammals have included Polecats, Pine Martens, Dormice and Badgers. He is a former Chairman of the Cumbria Wildlife Trust's Badger Group.